# SCIENCE
## WITHOUT
# LAWS

**SCIENCE AND ITS CONCEPTUAL FOUNDATIONS**

A SERIES EDITED BY DAVID L. HULL

# SCIENCE
## WITHOUT
# LAWS

RONALD N.
**GIERE**

THE UNIVERSITY OF CHICAGO PRESS

CHICAGO AND LONDON

Ronald N. Giere is professor of philosophy at the University of Minnesota and a former director of the Minnesota Center for Philosophy of Science. His books include *Understanding Scientific Reasoning* and *Explaining Science: A Cognitive Approach,* the latter published by the University of Chicago.

The University of Chicago Press, Chicago 60637
The University of Chicago Press, Ltd., London
© 1999 by The University of Chicago
All rights reserved. Published 1999
08 07 06 05 04 03 02 01 00 99    5 4 3 2 1

ISBN (cloth): 0-226-29208-8

Library of Congress Cataloging-in-Publication Data

Giere, Ronald N.
  Science without laws / Ronald N. Giere.
    p.   cm. — (Science and its conceptual foundations)
  ISBN 0-226-29208-8 (alk. paper)
  1. Science—Philosophy.  2. Realism.  I. Title.  II. Series.
Q175.G48898 1999
501—dc21
                                            98-46904
                                                CIP

∞ The paper used in this publication meets the minimum requirements of the American National Standard for Information Sciences—Permanence of Paper for Printed Library Materials, ANSI Z39.48–1992.

*For my mother,
Helen Agnes Marusa,
and in memory of my father,
Silas Irving Giere*

# Contents

| | | |
|---|---|---|
| Acknowledgments | | ix |
| Introduction | | |
| *The Science Wars in Perspective* | | 1 |

## PART ONE
### PERSPECTIVES ON SCIENCE STUDIES

| ONE | Viewing Science | 11 |
| TWO | Explaining Scientific Revolutions | 30 |
| THREE | Science and Technology Studies | 56 |

## PART TWO
### PERSPECTIVES ON SCIENCE

| FOUR | Naturalism and Realism | 69 |
| FIVE | Science without Laws of Nature | 84 |
| SIX | The Cognitive Structure of Scientific Theories | 97 |
| SEVEN | Visual Models and Scientific Judgment | 118 |

## PART THREE

## PERSPECTIVES ON THE PHILOSOPHY OF SCIENCE

|  | Introduction | 149 |
|---|---|---|
| EIGHT | Philosophy of Science Naturalized | 151 |
| NINE | Constructive Realism | 174 |
| TEN | The Feminism Question in the Philosophy of Science | 200 |
| ELEVEN | From *Wissenschaftliche Philosophie* to Philosophy of Science | 217 |
|  | Conclusion<br>*Underdetermination, Relativism, and Perspectival Realism* | 237 |
|  | Notes | 243 |
|  | References | 263 |
|  | Index | 281 |

# Acknowledgments

This volume owes much to numerous individuals and to several institutions. For general support and encouragement, I thank Paul and Patti Churchland, Arthur Fine, Bas van Fraassen, Jim Griesemer, Stephen Kellert, Philip Kitcher, Nancy Nersessian, Michael Ruse, Fred Suppe, and especially Paul Teller. I thank also my colleagues at the Minnesota Center for Philosophy of Science, John Beatty, Wade Savage, and Ken Waters. For help with my excursions into the history of Logical Empiricism I am indebted to Rick Creath, Michael Friedman, and Alan Richardson. Helen Longino, Lynn Nelson, and Naomi Scheman encouraged my interests in feminist theory. Tom Gieryn and Karin Knorr-Cetina continue to provide advice on current movements in the sociology of science. Special thanks go the series editor, David Hull, and to Susan Abrams, Executive Editor at the University of Chicago Press. Steve Lelchuk prepared most of the diagrams and helped in numerous other ways. Finally, of course, I am indebted to my wife, Barbara Hanawalt, for her encouragement in all things.

While preparing this manuscript, I enjoyed the support of the National Science Foundation and the National Endowment for the Humanities. The staff and Fellows of the National Humanities Center provided both intellectual stimulation and moral support throughout the 1997–98 academic year. I will remember them all with great affection for many years to come.

All but one of these essays (chapter 4, "Naturalism and Realism") have been previously published, and all but two within the past five years. For this presentation, they have been edited to remove overlaps and update references. No effort has been made to preserve their integrity as historical documents. I regard them as contributions to an ongoing contemporary debate. Here presented by permission, they first appeared as follows: Chapter 1, "Viewing Science," in *PSA 94*, vol. 2, *Proceedings of the 1994 Biennial Meeting of the Philosophy of Science Association*, ed. R. Burian and M. Forbes, 3–16 (East Lansing: The Philosophy of Science Association, 1995). Chapter 2, "Explaining Scientific Revolutions," in *Issues and Images in the Philosophy of Science*, ed. D. Ginev and R. S. Cohen, 63–86 (Boston: Kluwer, 1997). Chapter 3, "Science and Technology Studies: Prospects for an Enlightened Post-Modern Synthesis," *Science, Technology, and Human Values* 18:102–12 (1993). Chapter 5, "Science Without Laws of Nature," in *Laws of Nature*, ed. F. Weinert, 120–38 (New York: Walter de Gruyter, 1995). Chapter 6, "The Cognitive Structure of Scientific Theories," *Philosophy of Science* 61:276–96 (1994). Chapter 7, "Visual Models and Scientific Judgment," in *Picturing Knowledge: Historical and Philosophical Problems Concerning the Use of Art in Science*, ed. B. S. Baigrie, 269–302 (Toronto: University of Toronto Press, 1996). Chapter 8, "Philosophy of Science Naturalized," *Philosophy of Science* 52:331–56 (1985). Chapter 9, "Constructive Realism," in *Images of Science*, ed. P. M. Churchland and C. A. Hooker, 75–98 (Chicago: University of Chicago Press, 1985). Chapter 10, "The Feminism Question in the Philosophy of Science," in *Feminism, Science, and the Philosophy of Science*, ed. L. H. Nelson and J. Nelson, 3–15 (Boston: Kluwer, 1996). Chapter 11, "From *Wissenschaftliche Philosophie* to Philosophy of Science," in *Origins of Logical Empiricism*, Minnesota Studies in the Philosophy of Science, vol. XVI, ed. R. N. Giere and A. Richardson, 335–54 (Minneapolis: University of Minnesota Press, 1996).

# Introduction

*The Science Wars in Perspective*

The steadily growing influence of science and technology on all aspects of life will be a major theme in any retrospective assessment of the twentieth century. In spite of—and perhaps also because of—this unquestionable influence, the very nature of science and technology is, at the close of the century, a deeply contested issue. Indeed, so contested is the issue that the public media now regularly report on the latest battles in the so-called science wars. Increased publicity both reflects and reinforces increased polarization, with the result that there often seem to be only two sides. On the one side, we find what I would call "enlightenment rationalists" or "metaphysical realists," who are, however, often derisively referred to as "reductionists" or "essentialists." This camp includes most scientists, most members of the public, and some historians and philosophers of science. The other camp contains mostly intellectuals, some historians and philosophers of science, many sociologists of science, and many students of literature and culture more generally. To their enemies, these students of culture, scientific and otherwise, are merely "relativists" or "postmodernists." At their most polarized, the science wars are portrayed simply as battles between humanists and scientists.[1]

To be sure, one can find contemporary students of scientific culture claiming that "there is no obligation upon anyone framing a view of the world to take account of what twentieth-century science has to say" (Pickering 1984, 413). My rejection of such views will become clear very shortly. Here I wish to concentrate on the flavor of the opposition. This comes across clearly in a commentary on his spoof of cultural studies (1996a) by the physicist, Alan Sokal. "Anyone who believes that the laws of physics are mere social conventions," Sokal writes, "is invited to try transgressing those conventions from the windows of my [twenty-first floor] apartment" (Sokal 1996b, 62). Now, anthropologists tell us that our pre-literate ancestors were well aware that falling from great heights can be deadly. Such knowledge is grounded in quite basic human, and even animal, experience. Indeed, some early humans apparently put this knowledge to good use, securing food and hides by driving large animals over cliffs. But no student of language or culture would be so foolish as to claim that these ancestors had any knowledge of the laws of physics. Indeed, they could not even have had the *concept* of a *law of nature*. That concept, in its modern form, did not come to the fore until maybe as late as the sixteenth or seventeenth century in Western Europe. It seems not to have existed in China, which by then already had a very long tradition of philosophies of nature. Indeed, the most elementary histories of science inform us that Newton's conception of universal gravitation was initially greeted by many of the most intelligent people of his time with considerable skepticism. How, they asked, could one body possibly act at a distance on another? Later physicists introduced the concept of a gravitational field, which at least could be thought to act where the action is. Gravity and fields of force may not be "mere social conventions," but even to think in such terms requires fairly sophisticated concepts created relatively recently by people living in literate cultures. Part of understanding what science is all about is understanding how these humanly created concepts enable us to connect in such exquisite ways with the real world. The science wars distract us from pursuing such understanding.

Issues underlying what are now called the science wars have been debated for more than a decade within the science studies community itself. The essays collected in this volume provide one participant's perspective on these issues. It is not a perspective from inside either of the recently constructed warring camps. It is from between, beyond, outside, above, or below these positions—but it is not a view from nowhere. My *intellectual roots* are in the sciences, first as a student of physics, but later also as a follower of work in statistics, biology, geology, and the cognitive sciences. My *professional identification* has always been with the philosophy of science. Nevertheless, I have consistently been engaged, both intellectually and professionally, with neighboring disciplines, first the history of science, and later also the sociology of science and science studies more generally. My perspective is not *inter*disciplinary but, insofar as this is possible for a single person, *multi*disciplinary.[2]

I begin with the conviction that there exists much genuine scientific knowledge. Moreover, I firmly believe there have been dramatic increases in scientific knowledge during the twentieth century, particularly since World War II. We have learned, for example, that inheritance is carried by DNA molecules with a two-strand helical structure. And even the continents have not always been where they now are. In stating these convictions I am not merely playing games with the meaning of expressions like "scientific knowledge." I intend such expressions in the relatively ordinary sense that scientific knowledge is knowledge *of the world* and that there is a difference between knowledge and mere opinion, even widespread opinion.

Thus understood, most scientists and knowledgeable members of the general public would agree with these convictions. So what is the problem? It is that the more theoretical ways these simple convictions have generally been understood no longer square with other things that have been learned about what constitutes scientific knowledge and how it is acquired. The problem is not with current scientific theories of the world, but with current theories (or metatheories) of what it is to acquire good scientific theories of the world. As is typically the case for individuals, our collective self-knowledge lags behind our collective knowledge of the world.

The theories *about* science prominent in Europe and North America until around 1960 derived ultimately from the Enlightenment of the eighteenth century, a period in which the achievements of the seventeenth-century scientific revolution became incorporated into a more general cultural world view. A number of concepts came into prominence in particular ways during that time, including those of *laws of nature, scientific truth,* and *scientific rationality.* These particular concepts were connected, of course, in that it is through the power of rational inquiry that true laws of nature were thought to be discoverable at all. In spite of a romantic reaction in the early nineteenth century, later prominent dissenters such as Nietzsche, and some disillusionment produced by World War I, the Enlightenment picture of science seems to have dominated thinking about science from the mid-nineteenth to the mid-twentieth century. Indeed, this picture of science was reinforced in the twentieth century, inspired by the new physics, quantum mechanics and relativity theory, associated above all with Einstein. Some of the philosophers who founded what became known as Logical Empiricism explicitly thought of themselves as part of a "Second Enlightenment."[3] Fundamental concepts like those of laws of nature, scientific truth, and scientific rationality were retained, but reformulated in the idiom of the formal logic first developed by Russell and Whitehead around 1910. Laws of nature came to be understood as true universal generalizations and rationality as logical inference, above all deductive, but also, programmatically at least, inductive. The spirit of Logical Empiricism was not confined to philosophers of science. Following World War II, it spread, especially in North America, to the sciences themselves, particularly the behavioral and social sciences, and somewhat into the culture at large.

The Enlightenment picture of science was explicitly challenged in the early 1960s, particularly in Thomas Kuhn's *Structure of Scientific Revolutions.* The main lesson I take from Kuhn and many other students of scientific practice is that the very categories in which the Enlightenment view of science was formulated are inadequate to capture the actual practice of science, both historically and in its contemporary forms. Concepts like those of laws of nature, truth, and rationality are not givens, but are themselves interpretive categories which have their own histories.[4] Science need not be understood in these terms and, indeed, may be better understood in other terms.

This way of understanding the lessons of critical studies of science for the past generation explains part, maybe even a large part, of the intellectual basis of the current conflict over the status of science. Many people on both sides seem so to have internalized the Enlightenment view of science that, for them, to challenge aspects of that view is to challenge science itself, and, conversely, to defend science is to defend it in its Enlightenment form. From my point of view, neither of these reactions is correct. There are more general, or more basic, notions of scientific knowledge and progress that may be held apart from their Enlightenment presentations. The intellectual basis of the conflict, therefore, may be undercut by the realization that there are *other ways* of conceiving scientific knowledge.[5]

The underlying purpose of these essays is to develop and promote *one* other way of thinking about scientific knowledge. My deepest methodological commitment is to *naturalism*. This commitment is expressed not so much in terms of particular doctrines, but as a program—a way of approaching the subject. The primary methodological principle of naturalism, as I understand it, is not only to avoid appeals to anything supernatural, but also to avoid appeals to *a priori* claims of any sort. This principle is intended to have the consequence that any conclusions one reaches about the nature of science are subject to criticism based on theoretical, historical, psychological, or social investigations into particular scientific practices. My naturalism, therefore, rejects appeals to any presumed non-empirical philosophical method, such as logical or conceptual analysis, to reach conclusions about the nature of science.[6]

The fundamental concept in my particular understanding of scientific practice is that of a *model*. Models, for me, are the primary representational entities in science. Scientists, I claim, typically use models to represent aspects of the world. The class of scientific models includes physical scale models and diagrammatic representations, but the models of most interest are *theoretical* models. These are abstract objects, imaginary entities whose structure might or might not be *similar to* aspects of objects and processes in the real world. Scientists themselves are more likely to talk about the *fit* between their models and the world, a terminology I happily adopt.

I argue that what are called laws of nature function ambiguously in the actual practice of science. On the one hand, they may be regarded as *principles,* such as the explicitly named Principle of Covariance, the Uncertainty Principle, or the Principle of Natural Selection. As such, I claim, they are not even candidates for being truths about the world. They are not statements, but general rules for the construction of models. Incorporated into the characterization of particular models, however, they do function as true statements, but not as statements about the world. They are then truths only about an abstract model. In this context, such statements are true in the way explicit definitions are true. The empirical question—the question of realism—is how well the resulting model fits the intended aspects of the real world. And here my central claim is that the fit is always partial and imperfect. There is no such thing as a perfect model, complete in all details. That does not, however, prevent models from providing us with deep and useful insights into the workings of the natural world.[7] Employing concepts which frame contemporary professional debates, my view is that one can have *realism without truth.*

Somewhat contrary to the deliberately provocative title of this book, therefore, my view is not that there are *no* roles for the concepts of laws of nature and truth in a proper understanding of science. It is, rather, that those roles have been misrepresented in the Enlightenment view of science. Science does not deliver to us universal truths underlying all natural phenomena; but it does provide models of reality possessing various degrees of scope and accuracy. That dual conclusion may not provide extremists with all they desire, but I think it provides all that anyone can reasonably ask.

I hold a similar view regarding presumed canons of scientific rationality. Of course scientists make judgments regarding the fit between particular models and aspects of reality. And of course I do not claim that scientists' judgments in these matters are "irrational." It is the conception of rationality that is at issue. The original Enlightenment dream was of universal, categorical principles of rationality that could guarantee the truth of natural laws meeting their conditions. By the twentieth century, this dream was generally reduced to a desire for universal principles that yielded the *probable* truth of natural laws. My view is that there are no such principles. One can have *scientific judgment without rationality.*

Coming to hold that one model fits better than others is not a matter of pure reasoning or logical inference. Rather, it is a matter of making a *decision*. Effective decision making requires strategies for reaching desired goals. This applies to business and military decisions as well as to scientific decisions. If one wishes to talk of rationality here, it is a *conditional* or *instrumental* rationality, a matter of using effective means for reaching desired goals. In the case at issue, one goal shared by most scientists is to choose from among the available alternatives the model that best fits the real world. But other goals, such as a desire for personal or professional gain, cannot be excluded. Experimentation, however, provides an aid to making such decisions. Sometimes a fortunate combination of data and experimental design will lead all but the most stubborn of scientists to make the same judgment. That is as close as science ever gets to the Enlightenment ideal. In less ideal cases, a scientific consensus may rest more on shared values than on empirical data. It is not surprising, therefore, that both sides in the science wars can cite cases in the history of science supporting their positions. Both sides, I think, are guilty of selective attention and overgeneralization.

As an aid to the reader, I have grouped the chapters more by style and intended audience than by specific topics. The essays in part 1 were all written as overviews for audiences ranging from general philosophers of science to interested scholars in science studies and in the sciences themselves. They all situate my view of science among the current alternative views. These essays should be accessible also to people new to the field of science studies, perhaps even to advanced undergraduates. The essays in part 2 develop various aspects of my own view for an audience that includes scholars in all the science studies disciplines. The first of these, "Naturalism and Realism," links the general program of naturalism with a version of scientific realism that is both constructive and perspectival while remaining genuinely representational. The other essays in this group develop aspects of the view that it is models, not laws in the traditional sense, that do the important representational work in science, and that judging the fit of a model to the world is a matter of decision, not logical inference. Finally, the essays in part 3 are directed more specifically to my fellow philosophers of science, although even these should be accessible to members of a wider science studies community. The final chapter, in particular, is a prolegomenon to the historical study of Logical Empiricism in North America. It raises questions and suggests hypotheses that future historians of the philosophy of science from roughly 1938 to 1962 might find useful. Above all, it suggests that the philosophy of science itself exhibits the deep historical contingency that, from my perspective, is characteristic of all scientific knowledge.

# PART ONE

## Perspectives on Science Studies

# ONE

# Viewing Science

## 1. Introduction

In this chapter I will be focusing on those of us engaged in viewing science, particularly philosophers, psychologists, historians, and sociologists of science. I will be doing the viewing from the vantage point of the philosophy of science, which thus fills the foreground. The middle distance will be occupied by the sociology of science, while the history of science and cognitive studies of science occupy the background.

## 2. Philosophy of Science in Historical Perspective

I begin with a historical view of the philosophy of science itself. The most common picture of the recent history of the philosophy of science in North America is that, after a long period of dominance, Logical Empiricism was superseded in the 1960s by a historical approach to the philosophy of science inspired by Kuhn's (1962) *Structure of Scientific Revolutions*. That picture presents a very distorted view of the historical landscape. Nevertheless, there have been some large issues at stake since the 1960s. One large issue is this: How are we to understand the practice of the philosophy of science itself? In particular, how are we to understand relations between the philosophy of science and the history, psychology, and sociology of science? I suspect that part of the reason the contrast between logical and historical approaches to the philosophy of science has seemed so important to so many is that it implicitly raises these reflexive questions.

These questions are not new. They were explicitly debated both in the United States and in Germany in the 1920s and 1930s. John Dewey provides an exemplar of how the debate went in the United States. By 1929, when he turned seventy, Dewey was a philosophical naturalist, and to some extent even an evolutionary naturalist. That is, he rejected all claims to knowledge of the world based on anything but empirical scientific methods. There was, for Dewey, no special philosophical knowledge, particularly none that could provide a foundation or ultimate legitimization for the sciences. Rather, our understanding of evolutionary biology, of psychology, and of culture provides a basis for an understanding of scientific inquiry itself. What was special to philosophy, for Dewey, was the modernist task of bringing to moral and political inquiries the conclusions and methods of the sciences. He was far less concerned with the truth of scientific conclusions than with their usefulness for solving current societal problems.[1]

My exemplar of the debate in Germany focuses, of course, on a small community of philosophers and scientists, operating on the fringes of the German philosophical world, advocating a "scientific philosophy." The history of this movement is only now emerging from the realm of disciplinary founder myths into that of historical scholarship. Here I draw only briefly, and very imperfectly, on this new scholarship.[2]

In retrospect, one could describe the program of scientific philosophy as that of "naturalizing" philosophy, where philosophy is understood as consisting primarily of neo-Kantian metaphysics. More specifically, in scientific philosophy the problem of developing a *philosophical* understanding of arithmetic and geometry, and of space, time, and causality, was to be eliminated in favor of a *scientific* understanding of these concepts provided by the then new research into the foundations of arithmetic and geometry, and by the new physics, particularly relativity theory and quantum mechanics. But if philosophical metaphysics is replaced by scientific theory, what role remains for the philosopher of science? Carnap (1937) provided the canonical answer: Philosophy of science becomes the logical analysis of the language, concepts, and theories of the sciences, an enterprise that, like modern mathematical logic itself, takes place in the philosophically secure realm of the analytic *a priori*. It is this view of the philosophy of science that underlay what became Logical Empiricism in post–World War II America.

For Dewey, as noted above, the methods of empirical science were themselves a subject for scientific inquiry drawing heavily on psychology. Prior to 1933, the German scientific philosophers had only passing interests in empiricist methods as such. They were more concerned with the general possibility of an objective correspondence between the structure of experience, or of language, and the structure of the world—a typical Kantian concern. As they moved into the Anglo-American context, the logical empiricists had to take a stand on more traditionally empiricist questions about scientific method. They decided that such questions belonged in the province of philosophy, and are thus a matter of logic rather than psychology or any other science. This decision was codified in the famous distinction between "the context of discovery" and "the context of justification."

Ironically, most of the later critical literature on this distinction implicitly honors it by considering only its legitimacy and not inquiring into its origins.[3] The recognized source of the distinction is Hans Reichenbach's *Experience and Prediction,* published by the University of Chicago Press in 1938.[4] This book was written, in English, during the years 1933–38 at the University of Istanbul where Reichenbach, along with fifty or so other former German professors, found refuge in Mustafa Kemal's new Republic. He had been dismissed from his post as professor of the epistemology of the natural sciences in the physics (not philosophy!) department at the University of Berlin following imposition of Nazi racial laws in early 1933. Versions of the distinction had been common in German philosophy for more than fifty years. Reichenbach introduces his English version of the distinction in the very first section of *Experience and Prediction* where he sets out the task of epistemology as he conceives it. Here he is primarily concerned to distinguish epistemology from psychology. This he does by associating the concerns of epistemology with those of logic, and then drawing on a more recent tradition of distinguishing logic from psychology. Later in the book he cites the example of Einstein and the general theory of relativity, drawing an explicit contrast between "the man who found the theory" and a "[logical] relation of a theory to facts" (1938, 382). This contrast exactly parallels that between the contexts of discovery and justification.

I suggest that part of the significance of the distinction for Reichenbach at this time was its implicit denial that characteristics of a *person* proposing a scientific hypothesis have anything to do with the scientific validity of the hypothesis proposed. This applies, in particular, to that person's being a Jew. Reichenbach seems to have made it a precondition on any scientific epistemology that it rule out the possibility of any distinction between, for example, Jewish and Aryan science. But I think there was more to it than this. Separating questions of the origins of ideas from questions of their validity seems to have been for Reichenbach, at that time, a matter as deeply personal as it was philosophical. And this sentiment surely must have been shared by everyone in the movement.

If one is going to insist on so strong a distinction between discovery and justification, one is obliged to produce a theory of justification to back it up. That Reichenbach did. His own theory of induction does satisfy the precondition that the justification of a hypothesis be independent of its origin. His rule of induction operates as a relationship between purely formal aspects of a fixed set of data and a single hypothesis—a relative frequency in a finite sequence of occurrences and a postulated limiting relative frequency, respectively. There is simply no place in such a formal relationship for any aspects of the wider context to enter into the calculation. This is not to say that Reichenbach, as well as contemporaries like Carnap and Popper, did not have other motives for wanting to make empirical justification a purely formal relationship. But I do think the particular historical circumstances at that time strongly reinforced those motives.

So how did a dissident European movement advocating the

replacement of much established philosophy by a new scientific philosophy transform itself into an establishment North American philosophy of science? And how did a naturalistic pragmatism incorporating an empirical theory of inquiry get replaced by a philosophy that reduced the philosophical study of scientific inquiry to the analysis of a formal relationship between evidence and hypothesis? That episode is only beginning to be subjected to historical study.[5] One can only wonder how the philosophy of science in North America would appear today if the Social Democrats rather than the National Socialists had come to power in Germany in 1933.

At this point I can imagine some of my philosophical colleagues objecting that the history of Logical Empiricism, how it came to prominence in North America, and how it changed in the process, is quite irrelevant to the question of its merits as a philosophy of science. The fact that such an objection can still resonate strongly within the community of philosophers of science demonstrates the enduring power of Reichenbach's distinction and the continuing reluctance to see Logical Empiricism as the contingent historical development it surely must have been. But we must learn to see it as such if we are ever to put aside old arguments and devote our undivided energies to developing new views of the nature of science.

The deep commitment to Reichenbach's distinction following World War II helps to explain the initial very negative reaction to Kuhn's work among philosophers of science. For Kuhn's view was that the history of science is a story of major changes in accepted theory driven by historical contingencies: by socialization into a scientific specialty, by psychological crises following the failure of established techniques to solve recognized problems, by sudden gestalt switches, by older opponents dying, and by textbooks being rewritten. Here there is no formal logical relationship between data and theory to tell us which theory is the better justified, and thus no way to separate origins from validation. Indeed, there is no such thing as validation in the older sense. Kuhn was, of course, accused of relativism. Motivating this charge, though seldom even alluded to in print, was, I think, the specter of Jewish science.

Contrary to common historiography, the predominant reaction to Kuhn among analytically trained philosophers of science was to forge closer links not with the history of science, but with contemporary science itself. This was particularly evident in territories already much explored by the logical empiricists—physics, and probability and induction—but also in the newly rediscovered realm of the philosophy of biology.[6] Viewed in retrospect, the form this reaction took now seems to me to have been profoundly mistaken. Abandoning the logical empiricists' clear distinction between a science and the philosophical analysis of that science, many philosophers of science in the 1960s and 1970s came to view their work as *continuous* with that of scientists in the fields they studied—physicists, statisticians, and biologists. To think that people trained in logic and philosophy should actually contribute to the solution of major theoretical problems in the sciences sounds presumptuous. Mainly it was naive. The enterprise assumes, quite mistakenly, that one can extract the theories of a science from their disciplinary culture and analyze them in the abstract. Later analytic philosophers of science were thus victims of an assumption they adopted uncritically from their logical empiricist elders. The typical result has been the creation of relatively isolated subdisciplines populated by philosophers and a few scientific sympathizers. The sciences in question have continued to develop following their own dynamics.

But those philosophers of science who turned to history did not necessarily choose a better path. To the extent that there has been a "historical school" in the philosophy of science, Kuhn's inclusion in that school has been problematical. Steven Toulmin (1972), Ernan McMullin (1970), Dudley Shapere (1984), Imre Lakatos (1970), and Larry Laudan (1977), to name only a few of the most prominent candidates, have been as united in their opposition to Kuhn as in their opposition to Logical Empiricism. Their joint project, pursued in different ways, has been to show that science exhibits *rational progress*. So the project became one of offering an account of scientific rationality. But what is the status of any such account? Or, what is the conception of the project of philosophy of science in which such an account is offered? For the most part, it seems to me, the project has been one of analytic philosophy in historical robes—providing a conceptual analysis of rationality using historical rather than logical categories. Among those mentioned, only Lakatos (1971) and Laudan squarely faced up to this metamethodological problem, and only Laudan (1987) seems finally to have broken with the analytic tradition to embrace a form of naturalism.

I would now like briefly to explore the extent to which accounts of rational progress offered by members of the historical school succeed in preserving a separation between discovery and justification sufficient to rule out the relativism they perceived in Kuhn. But first let me swing our view around to another direction. Sandra Harding (1986) has popularized "the science question in feminism" for feminists. There are equally serious "feminism questions in science" for philosophers of science. For example: Is it possible that the actual *content* of methodologically acceptable science might reflect the specifically gendered interests of the predominantly male scientists who created it? A positive answer to this question follows directly from an assumption which many members of the historical school quite correctly borrowed from Kuhn, namely, that theory choice always involves a comparative evaluation among existing rivals which do not exhaust the range of logically possible rivals. This assumption is clearly built into both Lakatos's and Laudan's accounts of rational progress as arising out of a clash among rival research traditions.

The argument is simple. Suppose there are two rival theoretical programs which together do not exhaust the logically possible programs. Suppose one program ends up more progressive by the stipulated methodological criteria. That it does so depends on which among the possible rivals was in fact the actual rival. Against other logically possible rivals, the current favorite might not have fared so well. So we must consider the process by which possible rivals get into the game. There is nothing in any of the proposed accounts of rational progress to prevent gendered interests, or any other sort of interest, from playing a major role in this process. It is simply a matter of persuading enough investigators to consider a possible rival as a serious alternative. And who can deny that gendered interests are powerful persuaders? Thus any sort of theory exhibiting only moderate success by the stipulated criteria can end up as the progressive choice, provided only that its de facto rivals do significantly worse.

And what goes for gendered interests goes for religious interests, or any other sort of interest. The possibility of Jewish science is not eliminated in these accounts of rational progress.

I think we will just have to live with the empirical possibility that, at any given time, our best sciences may nevertheless embody all manner of cultural interests and values in their very content. But, like feminist empiricists, I also believe that particular interests embedded in specific theories can be identified, and sometimes eliminated, by creating empirically superior theories.[7] So my view of science differs from those for whom the inclusion of cultural interests within the content of science is not subject to empirical criticism.

## 3. A View of Constructivist Sociology of Science

Among those willing to treat spoon-bending on a par with traditional establishment sciences (Collins and Pinch 1982), the possibility that the content of any science might be influenced by interests based on religious tradition or gender could hardly be denied. Nor are there strong motivations to deny it. I have suggested that upholding a strong distinction between origins and validity was emotionally rooted in the personal experiences of the founders of Logical Empiricism during the 1930s. For the founders of constructivist sociology of science, by contrast, the formative experiences were those of the 1960s. In Europe these experiences included not only the Vietnam war, but also Prague Spring and the student revolts. Here science was seen not as a savior, but as a villain, part of the established authority to be resisted. The project became one of critique, indeed, of undermining the claims of the sciences to any special cognitive authority. Thus, as I see it, a significant source of the current antagonism between philosophy and sociology of science reflects the different experiences of different generations—roughly the generation of the 1930s versus that of 1960s. That is a difference in viewpoints unlikely to be reconciled by verbal argument. For those for whom neither the 1930s nor the 1960s have been particularly formative, I offer the following pictures for your viewing pleasure.

First, it is worth remembering that there has been generational conflict within the sociology of science as well as between sociologists and philosophers of science. Robert Merton's work, which dominated the sociology of science up to the 1960s, was grounded in the experiences of the 1930s and 1940s. The essay introducing his famous four norms constituting the ethos of science, which now goes under the title "The Normative Structure of Science" (Merton 1973), was originally published in 1942 under the title "A Note on Science and Democracy," and reprinted in 1949 under the title "Science and Democratic Social Structure." The essay seems designed to exhibit a correspondence between the ideals of science and the ideals of liberal democracy. Certainly the specter of Nazi Germany looms large in this early essay. And the methodological autonomy of science is never questioned.[8]

Returning to later developments, there has never been a single view with the title "social constructivism." At the moment there are maybe a half dozen distinct views that could claim the title. In order to avoid too long a show, I will introduce a couple of philosophical filters, as it were, to reduce the multiplicity of views to two. That simplifies the overall picture, of course, but leaves it rich enough for my purposes.

The one view I will label *epistemological* constructivism. This view is explicitly agnostic about the existence of the entities and processes reported by scientists. Such things may or may not be there in the world independent of any social practices. But, as a matter of fact, if one looks carefully enough at the actual historical sequence of events through which scientists come to hold the beliefs they do, one finds the major determinants of their beliefs in the realm of social interests, interactions, and associations. The influence of the supposed entities on actual beliefs, even if these entities were to exist as claimed, turns out to be minimal at best.[9]

The second view I will label *ontological* constructivism. On this view the entities and processes named in theoretical scientific discussions are *constituted* by the social practices, interactions, and associations of scientists. Like positive laws in the Anglo-American system of justice, it makes no sense to abstract from the social context and speak of scientific laws as if they existed antecedently to, or apart from, the social context.[10]

My philosophical filters remove most of what is exciting about constructivist studies. But the excitement comes mainly from the richness of the cases, not the theoretical coloration. The theoretical views suffer from a major anomaly that has been there from the start, and has been recognized as such. It just has never been resolved. This anomaly is the problem of reflexivity, and it exists for both epistemological and ontological views of constructivism. The problem arises if we ask: What is the character of constructivist sociology of science itself? The problem becomes a reflexive problem if one answers that the sociology of science is itself a science, subject to the same sorts of investigation constructivists have carried out for other sciences. Epistemological constructivists would have to conclude that their own beliefs about the scientists they study were determined more by their own interests and social interactions as sociologists than by whatever might really be going on among their subjects. For ontological constructivists, the results of such an investigation would have to be that the objects of investigation, the beliefs, interests, etc., of their subject scientists are constituted by their own practices as constructivist sociologists.

The initial response to this realization was simply to accept it. That is not an inconsistent position. Of course it appears self-defeating from a realist perspective in which scientific investigation is supposed to yield at least a tolerably good picture of what is really going on. But if one's goal is radical critique, what better way to exhibit the total futility of any attempt to establish scientific authority? Thus, as I read him, Steve Woolgar now recommends that we celebrate the reflexive irony of total deconstruction, including self-deconstruction (Woolgar 1988b). Another response to reflexivity, which I associate with Harry Collins, is to be a realist about the social world, but a constructivist about the natural world (Collins and Yearley 1992). That position at least avoids self-deconstruction, but it requires a dubiously sharp demarcation between the social and natural worlds.

Of course there remains the possibility of denying reflexivity. This would remove the sociology of science from the realm of the sciences altogether. That is what the scientific philosophers did for philosophy of science by understanding it as logical analysis rather than as science. Perhaps this is part of the strategy of those who associate constructivism with hermeneutics or ethnomethodology.

Realism, by contrast, is immune to the disease of reflexivity. On a realist view, scientists sometimes succeed in discovering how some antecedently existing bit of the world actually works, and we in science studies, likewise, may sometimes succeed in explaining how they in fact did it. That explanation often involves exhibiting a causal link between the scientists and what they discover. Of course they don't always succeed, and neither do we; but we might sometimes succeed in explaining why they don't when they don't. In either case, our explanations of how we ourselves succeed are of the same type as our explanations of how they succeed.

Nor does realism undermine the project of critique—on the contrary. For illustration, take Andy Pickering's (1984) account of the weak neutral current in *Constructing Quarks*. Read from a realist perspective, which is how most scientists would read it, the book has real punch. A Nobel Prize was awarded for the discovery of the weak neutral current, and it is now part of the "standard model" of high-energy physics. But, assuming Pickering was writing as an epistemological constructivist, his conclusion was that the belief that such a phenomenon exists, and the whole practice of high-energy physics incorporating that belief, would have been adopted largely as it was *even if nature contained nothing resembling the supposed weak neutral current*. One need not accuse the scientists involved of deliberately engaging in deception, for they could be as deceived as everyone else. Nevertheless, from a realist perspective, if Pickering is right, this should be an international scandal. But such a conclusion depends on there being a workable distinction between cases in which there is a real discovery and cases in which the supposed objects of discovery are mere social constructs. And that distinction is denied by all forms of social constructivism.

In these latter comments I have invoked a realist vision of science, and realism is regarded as problematic not only by most recent sociologists of science, but also by many philosophers of science. So let me add some details to this vision. It will ultimately be a vision not only of the philosophy of science, but of science studies as a whole.

## 4. A Realist Vision of Science

A lot depends, as we have seen, on how one conceives the goals of one's discipline. Constructivist sociologists of science begin with the assumption that what needs to be accounted for are the commonly held beliefs of members of scientific communities. But how the world might be constituted is thought to have no direct influence on anyone's beliefs. It is only *other beliefs* about the world that matter. There seems, therefore, no way logically to force consideration of anything beyond beliefs. So the social study of science can proceed in complete autonomy from what might be the character of the world beyond anyone's beliefs.

For philosophers, the logical structure of this situation is all too familiar. Classical phenomenalism, for example, begins with the assumption that all we ever really experience are our own sensations. From that assumption flows "the problem of the external world." Similarly, empiricists of less radical persuasions (van Fraassen 1980) begin with the assumption that all evidence consists of what is observable, and then challenge anyone logically to force them to move beyond claims about the observable. These are fool's games because it is impossible to meet the challenge as formulated. They present, as John Dewey once said, problems not to be solved, but to be gotten over. So the following views may be regarded as designed to help one get over the problems caused by isolating science studies from the internal content of the sciences it investigates.

What should then be the goals of science studies? They are, of course, numerous. Most generally, we should seek to provide our various cultures with an understanding of how different sciences operate within those cultures. This requires developing a set of concepts in terms of which all of us, including scientists, can understand the development and workings of the various sciences. And this understanding should sometimes provide a basis for criticizing both particular practices in science and specific uses of science. But included in this understanding, I would insist, must be some explanation of how the social practice of science came to produce, and continues to produce, the extraordinary range of knowledge it does in fact produce. In short, one of the goals of science studies should be to explain the successes of modern science.

I hasten to add that by "success" here I mean what most people outside the philosophy or sociology of science mean by success. We have learned that the Earth revolves around the Sun, not vice versa. We now know that there are not only five planets, but at least nine. We know that there are atoms, and that the speed of light is finite. We know that the continents move and that DNA has two strands. These things are now known, and not merely socially acceptable to believe. Of course it is logically possible that even such robust claims as these might turn out to be mistaken, and it is logically possible to question any one of them. But that only means that our knowledge is not logically infallible, not that we don't really know these things. That it seems necessary to highlight such obvious examples of scientific success should be regarded as a scandal in science studies.

"How naive can you be?" I imagine a critic saying. If you were a theologian in 1492, you would be telling us how no one could possibly deny the divinity of Christ or the virgin birth. Of course that is right. But it is beside the point because the situations are not symmetrical. We have a big advantage over our fifteenth century counterparts, namely, five hundred years of hindsight regarding the historical development of science. The inability of anyone to collapse that hindsight into a logically compelling five-line argument is no basis for anti-realism, or even for agnosticism. On the other hand, taking a commonsense realist position regarding the history of modern science does not commit one to any of the myriad *interpretations* of scientific practice that have accompanied that history.

Take, for example, the notion of a universal law of nature, an idea associated with science at least since the seventeenth century, and still assumed in much of twentieth century philosophy of science. I have looked in vain for a broad-based historical treatment of this notion. From the bits and pieces available, I have concluded that the original view of science as discovering universal laws of nature had little basis in the actual practice of science, but was imported largely from theology. In this theology, God laid down the laws for human conduct and for nature. The task of natural philosophers, then, was to discover God's laws for nature, which are of course universal—except, perhaps, when God himself intervenes. In spite of its theological origins, the idea that there are universal laws of nature provided a powerful resource for Enlightenment philosophers. If the laws of the universe, both moral and natural, are discoverable by human reason alone, what need have we for priests and kings, and ultimately for God as well?

But one need not appeal to history to deconstruct the concept of a law of nature. The concept is theoretically suspect as well. For example, any law of nature refers to only a few physical quantities. Yet nature contains many quantities which often interact one with another, and there are few if any truly isolated systems. So there cannot be many systems in the real world that exactly satisfy any purported law of nature. Consequently, understood as general claims about the world, most purported laws of nature are in fact false. So we need a portrait of science that captures our everyday understanding of success without invoking laws of nature understood as true, universal generalizations.[11]

While we are being iconoclastic, what about the concept of truth itself? I don't mean the everyday concept which one might use to say "Yes, it is true that the continents move." I mean philosophical *theories* of truth such as a correspondence theory of truth. And not only theories of truth, but the whole of formal semantics which incorporates Tarski's version of the correspondence theory of truth. Should this view of truth, originally developed for investigations into the foundations of logic and mathematics, be part of our interpretation of science? I think not.

Though it was not his original intention, Hilary Putnam (1981) provided a reason for suspicion of this whole semantic apparatus nearly two decades ago. Suppose that God wrote down for us a complete description of the whole universe in the language of set theory, so that every statement therein was true. That, Putnam proved, in purely logical terms, would not uniquely fix the reference of the terms in those statements. In plain English, we could be given the whole truth about the universe, as expressible in set theory, and still not know what we were talking about. Putnam, who took realism to be the view that we may take the terms of our best confirmed theories as genuinely referring to real objects, concluded that realism is mistaken.[12]

In response, philosophers such as Nancy Cartwright (1983) and Ian Hacking (1983) argued that we can have direct evidence for the objects of scientific inquiry through experimentation, without worrying about having true theories. The linguist George Lakoff (1987) drew the more radical conclusion that the whole logical machinery of truth and reference which frames Putnam's analysis fails to capture the role of language in empirical science—whatever might be its virtues in the realm of logic and mathematics. It is not realism we should abandon, but the semantic framework in which Putnam initially framed it.

But if we abandon standard analyses of truth and reference, along with the notion of a law of nature, what resources remain to express a useful notion of realism? What remains, I think, is a more general notion of *representation*. In place of the usual exemplar of linguistic representation such as "The cat is on the mat." I would suggest beginning with maps, e.g., a standard road map. Maps have many of the representational features we need for understanding how scientists represent the world. There is no such thing as a universal map. Neither does it make sense to question whether a map is true or false. The representational virtues of maps are different. A map may, for example, be more or less accurate, more or less detailed, of smaller or larger scale. Maps require a large background of human convention for their production and use. Without such they are no more than lines on paper. Nevertheless, maps do manage to correspond in various ways with the real world. Their representational powers can be attested by anyone who has used a map when traveling in unfamiliar territory.

Nor is the connection between road maps and representation in science at all far-fetched. Oceanographers map the ocean floors, astronomers map the heavens (star maps are centuries old), geneticists are busy mapping the human genome, and neuroscientists are mapping the mind (or at least the brain). The recent appearance of numerous articles on visual modes of representation in the science studies literature is evidence that historians, philosophers, and sociologists of science are finally becoming aware of how much of science has been done, and increasingly is being done, using pictorial and diagrammatic modes of representation. Of course some of this literature is cited in support of a constructivist picture of science, but it can equally well be viewed as supporting a more liberal notion of realism, something we might call "perspectival realism."

There is a metaphysics (a scientific metaphysics, of course) that goes along with a perspectival realism. Rather than thinking of the world as packaged in sets of objects sharing definite properties, think of it as indefinitely complex, exhibiting many qualities that at least appear to vary continuously. One might then construct maps that depict this world from various perspectives. In such a world even a fairly successful realistic science might well contain individual concepts and relationships inspired by religious or gendered interests. It is possible, therefore, that our currently acceptable scientific theories embody cultural values and nevertheless possess many genuinely representational virtues.

Here we have a way of combining what is valuable in both constructivism and realism, but it requires abandoning the universal applicability of either view. We can agree that scientific representations are socially constructed, but then we must also agree that some socially constructed representations can be discovered to provide a good picture of aspects of the world, while others are mere constructs with little genuine connection to the world. This compromise does not reduce the sociology of science to the sociology of error. Explanations of success and failure remain symmetrical in that both invoke the same sorts of activities on the part of scientists. In particular, there is no presumption that success is the product of rational deliberation while failure results from the intrusion of social factors. There is no need to introduce asymmetric notions such as rationality.

## 5. Rationality

Although they are often associated, even equated, realism and rationality are two very different things. My picture of science is realistic in the sense that it allows for genuine correspondences between the natural world and the various representational devices deployed by scientists. Yet rationality does not appear in this picture—at least not the sort of rationality one finds in the writings of Aristotle, or in Enlightenment glorifications of science, or in the writings of many historically minded philosophers of science. That is primarily what makes my picture "naturalistic."

For Aristotle, rationality was an essential characteristic of humanity. Humans were defined as rational animals. In the Christian era, rationality became a property of the soul, thus making it possible later to argue the legitimacy of subjugating African slaves, and women, on the grounds that they lacked a rational soul. Since Darwin, however, the idea that there are essential characteristics of anything but abstract entities has been under constant attack. Humans, we now think, are born with capacities to develop various cognitive and sensory-motor abilities, but there are no genes for rationality over and above the genes that determine natural cognitive potentialities. So what can now be meant by attributing rationality to a scientist?

The only remaining legitimate use for the concept of rationality, I think, is in discussing the effective use of appropriate means to achieve desired goals. Then we can label as "irrational" people who employ manifestly inappropriate means in attempts to attain their goals. But this labeling has little to do with understanding how science is done. What is important are the goals of various scientific inquiries and the appropriateness of the methods used to achieve them.

Take, for example, the goal of determining whether antioxidants, such as Vitamin E, prevent heart attacks. We know that randomized clinical trials are more reliable than prospective studies for this purpose because they are more effective at eliminating alternative causal explanations for observed differences in the incidence of heart attacks in different groups of subjects. In general, the desirable characteristics of scientific methods are things like reliability, discrimination, efficiency, sensitivity, and robustness. Once one has determined the superiority of a given method in these terms, nothing of substance is added by labeling it more rational.

In rationality we have another concept that is long overdue for a sustained historical deconstruction. What historical research will reveal, I suspect, is a social construct that has served various interests in various historical contexts. In this vein, one wonders what interests have been served by the vigorous philosophical defenses of rationality and rational progress developed since the 1960s. My suspicion is that among the interests served has been the autonomy of the philosophy of science itself. Scientists investigate the effectiveness of various experimental techniques; historians record the progress made; but only philosophers get to pronounce these as rational or not. Ironically, many philosophers of science have been attacking social constructivism in the sociology of science in the name of their own social construct: rationality.

## 6. A Vision of Science Studies

Debates across disciplinary divides bear a discouraging resemblance to debates in politics and other areas of the public arena. Opposing views are reduced to simple stereotypes, and the point of debate is too often not understanding the subject matter, and certainly not understanding the opposing views, but maintaining one's disciplinary integrity. Thus one finds philosophers of science describing social constructivism as the view that "scientists *produce* the world" (Roth and Barrett 1990, 591). A reverse stereotype is a naive rationalism which portrays "scientific knowledge as the simple result of human rationality's encounter with reality" (MacKenzie 1990, 342). A vision of science studies that overcomes such stereotypes could produce a far richer picture of science than anyone now possesses.

I have already said that one of our goals should be understanding how science succeeds, when it does. This is admittedly contentious, though I have left open many possible different accounts of what constitutes success and how it is achieved. Less contentious is the view of science as a highly complex activity, at least as complex as the reality it investigates. This great complexity implies, I think, that it is impossible to obtain an adequate overall picture of science from any one disciplinary perspective. Different perspectives highlight different aspects while ignoring others. The only adequate overall pictures will be collages of pictures from various perspectives. There can be no uniquely adequate collage, but some may be put together with more skill and sensitivity than others, and may therefore be more enlightening than others—at least for some purposes.

In my picture, science studies cannot be autonomous. Its inquiries must draw on knowledge from many disciplines, including some of the sciences it studies. But is there a special role for the philosopher of science in this enterprise? Yes and no. Current training in the philosophy of science prepares one to be a good synthesizer of scientific knowledge and to use this knowledge in constructing theoretical models of various aspects of science. It also prepares one to raise normative issues, that is, to be critical of specific scientific projects in light of acquired empirical knowledge of how science typically works. And it prepares one to take part in debates over the role of science in the larger society. On the other hand, none of these activities are ones that could not also be carried out by historians or sociologists of science, or even scientists themselves, should they be so inclined. Just as philosophy can no longer claim to be queen of the sciences, neither can philosophy of science claim to be queen of science studies. But so long as we are willing to put aside our pretensions to possessing autonomous forms of knowledge, philosophers of science have much to contribute to the larger enterprise.

# TWO

# Explaining Scientific Revolutions

## 1. Science and the Philosophy of Science

It is difficult to exaggerate the importance of science and technology in contemporary culture. The popular media are full of reports on new developments in the biomedical sciences and in information and communication technologies, and on the role of science and technology in maintaining an economy that will be competitive in world markets. And these are only a few of the most prominent current areas of general interest. This widespread interest generates a need to understand science as a cultural phenomenon. What kind of activity is it? How does it work? How does it interact with other aspects of contemporary culture?

One might reasonably have expected philosophers of science to provide some enlightenment on these important issues. But if one turns to the literature within the philosophy of science, one may be disappointed. Many philosophers of science focus their energies on the logical or conceptual structure of particular scientific theories, such as relativity theory (Earman 1995) or quantum mechanics (Healey 1989). Even those philosophers of science whose interests transcend particular scientific theories tend to be preoccupied with the peculiarly philosophical concern to develop *normative* criteria for such activities as constructing scientific theories (van Fraassen 1980), formulating scientific explanations (Salmon 1984), or determining which of several rival hypotheses should be preferred (Howson and Urbach 1989). The source of these concerns is to be found not so much in the sciences as in the history of philosophy itself.

The Greek founders of Western philosophy were much concerned with the nature and sources of human knowledge. Later, in the Christian era, this concern was focused on knowledge of the Divine Will. Most philosophers were theologians, and they took it as their task to elaborate and defend the principles of Christian theology against heretics and other nonbelievers. As the impact of the Reformation took hold and the scientific revolution began to unfold, some philosophers took it upon themselves to defend the emerging new sciences, often against the skepticism of theologians. But their conception of the task, and their forms of argumentation, remained similar to those of the theologians they sought to displace. The assumed task was to provide an *autonomous,* philosophical, justification of the new science. Philosophical proofs of the justifiability of scientific knowledge have a similar overall structure to proofs of the existence of God. They begin with premises regarded as self-evident and proceed by logical demonstration to the desired conclusion.[1]

In the seventeenth century it was reasonable to assume both that science needed philosophical justification and that philosophy had within itself the resources to provide an autonomous, philosophical justification. As we near the end of the twentieth century, neither of these assumptions remains plausible. The obvious success of science, particularly in the twentieth century, eliminates any serious worry that there might be something fundamentally unsound about the way science is done or about its results. This does not mean that we understand why science has been so successful, or even that we are clear about what constitutes success in science, or about the connection between success and the veracity of the pictures of the world it has produced. But science now needs no wholesale philosophical legitimization. Indeed, the soundness of science as a whole is now more secure than any philosophical proof of its legitimacy could possibly be.[2]

As for the resources of philosophy to provide the kind of autonomous justification traditionally sought, these seem to have run out. If one denies oneself any appeal to empirical claims of a kind requiring scientific investigation, where is one to turn? To Logic? To intuition? To ordinary linguistic usage? These have all been tried without lasting success. That fact does not, of course, constitute a proof that no such demonstration is possible. But it does provide a powerful incentive to come down to earth and inquire, *within* a scientific framework, how science works and why it is successful. That, I suggest, is the real challenge to contemporary philosophy of science.

In what follows I will begin by reviewing post–World War II attempts to construct a general account of the nature of science. This review will include work by historians and sociologists of science, as well as by philosophers of science. One of the major features of more recent philosophy of science is its increasing involvement with these other fields, an involvement that has been both complementary and antagonistic. This review will provide the springboard for my own suggestions as to the directions future investigations should take.

## 2. Philosophy and Sociology of Science after World War II

From the end of World War II until roughly 1960, there existed two major theoretical approaches to the study of science as a human activity, one philosophical and one sociological. For the most part these approaches were *complementary.*

### Logical Empiricism

Until about 1960, philosophical thinking about science was dominated by Logical Empiricism. This philosophical movement was well named. It was *empiricist* in the classical sense that the ultimate appeal for all scientific claims was said to be direct sensory experience—and nothing else. It was *logical* in its methods, which derived from Russell and Whitehead's work on logic and the foundations of mathematics (1912–14), and from Russell's philosophical writings, particularly those on metaphysics and epistemology (1914). Two aspects of the logical empiricist analysis of science are relevant here.[3]

First, it was epistemological and highly "theory centric."[4] That is, it focused on the analysis of scientific knowledge, and it assumed this knowledge to be encapsulated in scientific theories. Moreover, it insisted that scientific theories are to be understood as axiomatic systems to which the methods of logical analysis could be applied.

Second, it insisted that the epistemological relationship between sense experience and theory is itself a "logical" relationship. Thus epistemology becomes a kind of applied logic. Of course it early became apparent that the "logic" could not be the deductive logic of mathematical reasoning. So the program was to develop an enriched *inductive* logic, of which probability logics were the most popular (Carnap 1950). The program of developing a probability logic for science continues to this day (Jeffrey 1985; Howson and Urbach 1989).

It follows that science has the properties of being both *representational* and *rational*. It is representational in the sense that theories are the kind of thing that can be *true* of the world. The probability of a theory is the probability that it is true. And science is rational because the probability of a theory provided by inductive logic was interpreted as expressing the rational degree of belief in the truth of that theory relative to existing evidence.

It is a consequence of the logical empiricist analysis of science that there is an unbridgeable gap between the content and methods of science and all other aspects, such as the psychology of scientists or their social organization. The former are fully analyzable using the *a priori* methods of modern mathematical logic. The latter are the subject of empirical social science. This gap was codified in the logical empiricists' rejection of "psychologism" and in their distinction between the "context of discovery" and the "context of justification."

*Functionalist Sociology of Science*

Between 1945 and 1960, the reigning sociology of science was the structural-functional approach of Robert K. Merton (1973). In this framework, the job of the sociologist was to exhibit the ways in which the *social structure* of an institution, as defined by its social norms, promotes its *function*. Here Merton was applying a general approach that was at the time used by sociologists in investigating a wide variety of institutions within society.

A nice example of structural-functional analysis is Merton's account of the role of priority disputes in science. They serve the function, he argued, of reinforcing the norm that original discoveries should be rewarded by suitable recognition. This is functional because the desire for recognition is an important motivating factor for scientists to pursue the goal of producing new knowledge.

Merton took it for granted that the function of science is to produce "certified knowledge." He explicitly renounced any role for the sociologist in analyzing the *content* of this knowledge or the *methods* by which it became certified. For Merton, then, the role of the sociologist of science is complementary to that of the logical empiricist philosopher of science. The philosopher is concerned with the logic of science, the sociologist with its social structure. These are distinct enterprises.

## 3. The Structure of Scientific Revolutions

During the past thirty-five years, Thomas Kuhn's *Structure of Scientific Revolutions* (1962) has emerged as the single most influential work on the nature of science to be published since World War II. The magnitude of its success seems to have surprised even its author. Part of its success was surely due to the timeliness and originality of its analysis. But there were fortuitous components as well. Like Darwin, whose *Origin of Species* was a quickly produced "outline" of a projected but unpublished larger work, Kuhn was also forced by circumstances to publish a condensed version of a projected longer work.[5] It is doubtful the larger work would have been so accessible to so many. In addition, by the middle 1960s, talk of "revolution" was very much in the air. Kuhn became a hero of the 1960s "cultural revolution" in spite of himself.

### Kuhn's Stage Theory

On the surface, Kuhn developed a "stage theory" of science. The stages in the life of a scientific field unfold chapter by chapter throughout the book. Indeed, the stages—pre-paradigm science, normal science, crisis, revolution, and new normal science—provide the main organizing theme of the whole work. Kuhn is at his best describing the characteristic activities of each stage. He is less successful at pinpointing the *mechanisms* that drive the process from one stage to the next. I will begin by following the order of Kuhn's own presentation, but my conclusion will be that, for the purpose of developing a general theory of science, the stages are misleading. It is the mechanisms that matter.

*Pre-paradigm Science*

Kuhn's major examples of pre-paradigm sciences are optics before Newton and electricity before Franklin. In such periods there is a roughly defined subject matter and two or more schools of thought about how the subject should be conceived. Typically these schools are social as well as intellectual entities, with disciples and students following the teachings of a single teacher. The different schools may coexist for a considerable length of time with little real contact among them and slow progress only within competing schools.

Even in Kuhn's own framework, "pre-paradigm" science is better described as *multi-paradigm* science. By his own characterizations of what constitutes *having a paradigm,* each of the competing schools has a paradigm. What is lacking at this stage are not paradigms, but a *dominant* paradigm that can guide the energies of the vast majority of practitioners concerned with the same general subject matter. Without a single, dominant paradigm, energy is wasted in fruitless debates over fundamentals. Only when such debate is past can all the practitioners focus their energies on new and often quite esoteric issues with some confidence that they are at least on the right track.

Thinking of this early stage as a multi-paradigm period actually strengthens the case for Kuhn's cyclical picture of science. It makes this early stage very much like the later "revolutionary" stage, which is also characterized by competition among fundamentally different approaches. And it reduces the question how a multi-paradigm situation resolves itself into a context with a single paradigm to the question of how a revolution gets resolved. The process should be fundamentally the same in both cases.

## Normal Science: The Meaning of "Paradigm"

The stage of normal science is characterized by a single, dominant approach to the subject matter. There is little questioning of the fundamentals of the approach. Rather, energy is devoted to working out its implications and extending its application to new phenomena. A phenomenon apparently at odds with the dominant approach is not taken as evidence that the approach might be mistaken. It merely creates a "puzzle" to be solved. How can this phenomenon best be incorporated into the existing framework? Failure reflects poorly not on the approach, but on the competence of the scientist.

*The Structure of Scientific Revolutions* is notorious for the ambiguities in its key analytical notion, "paradigm." In later writings (1977), Kuhn himself distinguished two quite different senses. In one sense, a paradigm is a very general "world view" that includes specific theories, instrumentation, and even metaphysical presuppositions. It is this very general sense of a paradigm as a "theory plus" that the initial philosophical commentators (Shapere 1964; Scheffler 1967) took as fundamental. This made it possible to assimilate Kuhn's problem of how revolutions get resolved to the standard philosophical problem of why one theory is to be preferred over another. Despite many of Kuhn's own comments, however, this was not the sense of "paradigm" that was in fact central to Kuhn's theory of science.

The sense that *was* central appeared in the early pages of *The Structure of Scientific Revolutions* where Kuhn emphasized the importance of "the concrete scientific achievement" (p. 11) which provides "models from which spring particular coherent traditions of scientific research" (p. 10). Thus, Newton's account of the motion of a planet or Franklin's account of the Leyden jar provided "paradigms" for the practice of mechanics and the science of electricity respectively. In later writings, Kuhn referred to these concrete achievements as "exemplars."

Contrary to a common view of scientific activity, normal science, for Kuhn, is *not* a process of "applying" general laws to new cases. Rather, solutions to new problems are developed by modeling them on the exemplary solutions that underlie the general approach. When Kuhn later talks about "the priority of paradigms" (ch. 5), he means that the practice of normal science is guided and sustained not by general theory or method, but by these exemplary solutions to earlier problems.

*Crisis*

At any point in a stage of normal science there will always be anomalies, that is, phenomena which no one has yet been able to explain with reference to standard exemplars. The tides, for example, were an anomaly for Newtonian mechanics for half a century. Since there are always some anomalies in any normal science research tradition, their mere existence provides no basis for dissatisfaction with the existing tradition. Nevertheless, there comes a time when practitioners in a tradition begin to lose faith that the anomalies can be resolved using the resources of their tradition. The result is a stage of *crisis*.

Just why scientists lose faith in their tradition is, as Kuhn admits (p. 82), very difficult to explain in general terms. It is clear that no simple answer in terms of the *number* of anomalies, or other qualitative features, can do justice to the historical facts. Nor, Kuhn argues, can it be explained by pressures from outside the immediate scientific community. So here we have a clear instance in which, as Kuhn himself suggests (pp. 85–86), the stage theory needs to be supplemented by deeper investigation of things like the psychological reactions and social responses of people when accepted ways of operating no longer seem to be paying off.

*Revolution*

Whatever its detailed causes, crisis leads to the proliferation of new approaches and a situation resembling "pre-paradigm" research. There is once again discussion of fundamentals and even of general methodological and philosophical principles. Eventually a new set of exemplars is created which gives rise to a new normal science tradition.

What leads a scientific community to abandon the old tradition for the new, to put aside one set of exemplars and take up another? Here Kuhn appeals to factors that are psychological, sociological, and institutional. Individuals, he says, experience something like a "gestalt switch." They come to see things differently, sometimes quite suddenly. Of course not everyone experiences a similar "conversion." However, those who do not convert tend to be members of the older generation who were educated in the old tradition. Eventually they retire or die off, leaving the field to a new generation that has been educated in the new tradition. Finally, new textbooks are written from the standpoint of the new tradition. Members of the old tradition are thus literally written out of the field.

*Incommensurability*

Why does Kuhn explain the resolution of revolutions by appealing to psychological, social, and institutional factors? Why cannot scientists simply compare two approaches and realize that one is objectively better? The reason, according to Kuhn, is that rival traditions are *incommensurable*. What does that mean?

In a great many passages, Kuhn explains incommensurability in *linguistic* categories. Adherents of different traditions, he says, talk past one another. They use the same words with different meanings. Moreover, Kuhn had a theoretical basis for this way of describing rival traditions. Wittgenstein taught that the meaning of words is a function of their use in actual life. Kuhn applied this idea to the scientific life.[6] And it fit in well with his emphasis on the importance of concrete exemplars of scientific achievement for defining a normal science research tradition.

Kuhn's insistence on the incommensurability of research traditions attracted considerable attention from philosophers. Prominent philosophers such as Kripke (1972) and Putnam (1975b) invented new theories of meaning and reference partly in response to Kuhn's apparent challenge to the objectivity of science. Nevertheless, in spite of Kuhn's own presentation, and its widespread acceptance by others, the linguistic interpretation of incommensurability seems to me mistaken. The important incommensurability is not one of meaning but of *standards* and *authority*.[7]

By Kuhn's own account, adherents of rival research traditions appeal to different exemplars to authenticate their solutions to new problems. Moreover, there are, according to Kuhn, no higher standards to which one might appeal. This is because the laws, theories, and methodological principles of a research tradition are themselves ultimately grounded in the exemplars. Exemplars are primary. Thus one need not invoke a dubious theory of linguistic meaning to conclude that a principled resolution of a conflict between rival research traditions is impossible. An incommensurability of standards is sufficient.

This understanding of incommensurability is very much in keeping with the *political* metaphor now associated with talk of "revolutions." The breakdown in a political revolution is not primarily linguistic. People may not listen to one another, but not because they cannot understand what is being said. The breakdown is one of *authority*. There being no court of ultimate appeal, one uses persuasion, manipulation, and, finally, force.

*Cognitive and Evolutionary Aspects of Kuhn's Account*

Although overtly a stage theory, there are indications in Kuhn's own work that the stages are not so fundamental as it may appear. One indication already noted is that Kuhn appeals to cognitive mechanisms, both psychological and social, to account for the dynamics of scientific development. For example, he claims that scientists do not learn or understand their subject matter in terms of general laws or theories. Nor do they proceed by following general methodological principles. Rather, they learn by example and appeal to these examples to validate solutions to new problems. Similarly, crisis occurs not because a general theory has been falsified, but because satisfactory solutions to new problems cannot be found. The crisis ends not because scientists believe some new theory to be true, but when sufficiently many practitioners in the field become convinced that some new solutions provide the most fruitful exemplars for future research.

From this more cognitive perspective, the stages appear as almost tautologous consequences of the simple fact that members of a scientific community do switch from one approach to another that is markedly different; for example, from a particulate to a wave theory of light. The main content of the stage theory, then, is the denial that scientific fields exhibit a cumulative development, and the contrary assertion that "radical" change does occur.

Most commentators on Kuhn's work have been aware that there is a cognitive theory underlying the stage theory. What is less often remarked is that there is also an *evolutionary* theory. It appears in the last few pages of *The Structure of Scientific Revolutions* where Kuhn attempts to explain how there can be scientific progress without there being any "truth" toward which science moves. His answer invokes an analogy with organic evolution. According to evolutionary theory, species do become better adapted to their environment, but there is no final form toward which they are evolving. Progress is measured relative to the previous state, not in terms of movement toward a final goal. Scientific progress, he claims, should be similarly conceived.

Kuhn warns his readers that "the analogy that relates the evolution of organisms to the evolution of scientific ideas can easily be pushed too far" (p. 171). And well he might, because to develop the evolutionary analogy would be to undermine the stage theory that informs everything that went before. The idea of stages itself has its roots in a *biological* analogy, that of the *development* of individual organisms—birth, growth, maturity, decline, death. But even when applied to other sorts of *individual* development, the analogy is weak, as in Piaget's (1954) stages of cognitive development or Kohlberg's (1973) stages of moral development. Applied to collective, social phenomena, it is even worse, as in the stages of economic and political development proposed both by Marx and by later capitalist thinkers. For social phenomena, *evolutionary* models may be much better, but then one loses any neat progression, or cycle, of "stages." Later I will suggest that we retain the cognitive and evolutionary aspects of Kuhn's account and abandon the stage theory altogether.

## 4. Post-Kuhnian Philosophy and Sociology of Science

The initial reaction to Kuhn's account, particularly by philosophers (Shapere 1964; Scheffler 1967), was very negative. The reason is not far to seek. Kuhn's account was neither representational nor rational. Moreover, he refused to recognize the barrier philosophers and sociologists of science had erected between the content and methods of science, and its social structure.

For Kuhn, science is not a search for a true representation of the world. It is a puzzle solving activity which results in something better characterized as an *interpretation* of the world. Furthermore, in appealing to psychological and sociological factors to explain the historical development of science, he implicitly rejected any search for rational principles of theory choice. Philosophers, in particular, saw Kuhn as opening the door to a dangerous *relativism*. If the choice of a research tradition is a matter not of logic but of individual judgment and social solidarity, what "objective" basis could there be for claiming that a new tradition is in any significant sense "better" than the old?

It is hardly surprising, therefore, that many philosophers and sociologists of science simply rejected Kuhn's account and went about their business as before. But others in both disciplines were more deeply influenced.

*Progress and Rationality*

Following publication of Kuhn's book there arose within the philosophy of science a "historical school" which took up the problem of accounting for the development of science. But by and large these philosophers retained the earlier philosophical goal of showing how the development of science could be, if not progressing toward the truth, at least objectively "rational."[8]

Laudan (1977, 1984) provides a good example of a philosopher deeply influenced by Kuhn's account, yet deeply critical of it. Like most philosophers, he focuses on the more global interpretation of "paradigm," reinterpreting Kuhn's "research tradition" as a series of theories, each understood as a set of laws, pretty much as portrayed by logical empiricists. Following Kuhn, he takes it as the business of a research tradition to "solve problems." But, unlike Kuhn, he thinks that the number and importance of problems solved within a tradition can be used as an objective measure of the "problem-solving effectiveness" of that tradition, and thus of the *acceptability* of the tradition and of its associated theories. He even suggests that the *rate* of problem-solving effectiveness can provide a measure of the pursuit-worthiness of a tradition. Objective *progress* is measured by problem-solving effectiveness. Scientific *rationality* is then characterized as accepting the more progressive of rival traditions while pursuing the more promising. The particular judgments of individual scientists play no role in the account. Nor is there any role for social or institutional variables.

In the end, therefore, Laudan claims to recover much of what philosophers and sociologists of science thought they had achieved before Kuhn came on the scene. He portrays science as a rational activity in which the content and methods of science are distinct from its social structure. The price he pays is giving up the picture of science as a representational activity in favor of Kuhn's account of science as a problem-solving activity.

Note that on Laudan's account there is no incommensurability among research traditions. Problem-solving effectiveness can be calculated within each tradition and the tallies compared. Thus there are no "scientific revolutions" in Kuhn's sense. There is just the rational choice by scientists to pursue and accept the objectively more promising and more acceptable traditions. As a result, some traditions fade away while others prosper.

None of the philosophical reactions to Kuhn's account has gained widespread support. In Laudan's case, for example, there are serious questions about its adequacy both as a *descriptive* account of how science has been done and as a *normative* account of how it should be done. Few historians or philosophers of science, including Laudan, have even attempted to show in detail that scientists in historically important cases actually made their choices of pursuit or acceptance in accord with standards based on problem-solving effectiveness.[9]

Nor is it clear that the supposed standards could be applied as Laudan intends. If judgments of problem-solving effectiveness were to be made by members of the rival traditions, it is pretty clear that members of each tradition would give their own tradition higher rankings. In particular, each would regard its exemplary problem solutions as being very weighty indeed. But if the scientists themselves cannot be trusted to produce an objective ranking, who else is to hand out the blue ribbons? And what difference would it make to the scientists themselves? Why should they pay attention to "outsiders"?

*Science as a Social Construct*

During the past several decades, European sociologists of science have developed an alternative to the functional approach of the Mertonian school. Their approach emphasizes Kuhn's notion of a "paradigm" as an *exemplar* for guiding further research. And, like Kuhn, they insist on the indispensability of judgments by scientists, both individually and collectively. The result is an account of science which, like Kuhn's, portrays science as being nonrepresentational, nonrational, and lacking any fundamental separation between content and methods, and social structure.

British sociologists, particularly the group at Edinburgh, have pursued a conclusion that Kuhn resisted. If it requires individual judgment to determine which are the exemplary solutions and what counts as a successful application to a new case, then there is room for what even Kuhn would regard as "nonscientific" values and interests to enter into the scientific process at the most fundamental level. These interests might be personal, professional, social, political, or a combination. Historians and sociologists associated with the Edinburgh school have produced an imposing set of historical cases which they claim illustrates the role of various kinds of interests in the development of science.[10]

Other "new wave" sociologists of science are even more radical. It is not just that nonscientific interests "influence" scientists' judgments about theories. Rather, science is totally *constituted* by human interests and interactions. Science is simply a social construct, like morals or the law. For these sociologists, then, moral relativism and scientific relativism are of a piece. Just as there is no objective basis for preferring our Western cultural values to those of other peoples, so also there is no objective basis for preferring our Western scientific belief system to theirs.[11]

*Summing Up*

One does not have to believe in any extrascientific basis for scientific rationality to be quite sure that, in some sense, the pursuit of science during the last three hundred years has produced a greater understanding of the physical world than was available earlier. And, indeed, progress in the past fifty or one hundred years has been dramatic. Yet, over thirty-five years after publication of *The Structure of Scientific Revolutions,* we are left with at least two very different approaches to understanding how science works. Philosophers, focusing on one sense of "paradigm," have been seeking objective criteria for rational scientific progress. Sociologists, focusing on another sense of "paradigm," argue that scientific progress is no different from political or social progress.

One wonders whether there is not a middle way. I think there is. It may be characterized by a different combination of the traditional scientific virtues than that embraced by either historically inclined philosophers of science or recent sociologists of science. The winning combination, I suggest, is one that gives up the search for criteria of scientific rationality, abandons the attempt to separate the content and methods of science from psychological and sociological reality, but preserves the view of science as a representational activity.[12] A framework for such an account may be found, I suggest, in the cognitive sciences and in evolutionary models of scientific fields. As noted earlier, these components of an account of science were already present in Kuhn's original work, but were largely ignored by his philosophical and sociological heirs.

## 5. Evolutionary Models of Science

There are several important negative lessons to be learned from the past thirty-five years of debate about "scientific revolutions." One is that we should abandon the *political* metaphor now embodied in the very term "revolution."[13] Thinking of scientific changes this way makes them seem more social, and more arbitrary, than they in fact are. It unnecessarily encourages the idea that scientific facts might be just like social facts. In science, there is not only interaction among individuals and social groups, there is also causal interaction with the world. Although it cannot be claimed that this causal interaction uniquely determines the scientist's pictures of the world, it does play a major role in the story. The task is to explain that role, not to ignore it or deny its existence.

A second negative lesson is finally to abandon the *developmental* metaphor that underlies both Kuhn's stage theory of science and its recent philosophical rivals. Given the generally low repute accorded stage theories of other social phenomena, it is surprising that Kuhn's stage theory of science has not been more severely criticized on these grounds alone. Stage theories take a biological concept that applies primarily to the development of *individual* organisms and attempts to apply it to social groups. Taking the evolutionary analogy more seriously provides an immediate explanation of why developmental models do not work very well at the collective level. The variety of possibilities for evolution in a population is very great. The actual path taken in the evolution of a biological population depends not only on variations in the gene pool, but also on myriads of details in the local environment. Much is left to chance. So there simply cannot be a single pattern of evolution for populations. The circumstances are far too variable.

In fact, we can now see why Kuhn's political metaphor and his stage theory are at odds with each other. Political revolutions, like social phenomena generally, are highly variable. Any stage theory of political change is immediately subject to many counterexamples which can be avoided only by rendering the theory so vague as to empty it of useful content. This "internal contradiction" in Kuhn's views has been obscured because "revolution" was only one stage in his theory. But unless one grossly exaggerates the differences between "normal" and "revolutionary" science, the political analogy must be applied to the whole theory, or none of it. The somewhat paradoxical conclusion is that to explain scientific "revolutions" one should first stop thinking of them as revolutions in the full-fledged political sense. Rather than trying to understand "scientific revolutions," we should be thinking in terms of "the evolution of science."

## The Evolution of Scientific Fields

There is currently no consensus on how best to proceed to develop an evolutionary model of science. Here I can only identify the major alternatives and indicate the directions that now seem to me most promising for further research.[14]

Evolutionary theory applies to *populations* of individuals. What, then, is the population to which an evolutionary model of science might apply? The literature contains two sorts of answer to this most fundamental of questions. The answer favored by philosophers and intellectual historians of science, such as Toulmin (1972), is a population of something conceptual, such as theories or concepts. On this approach one investigates the evolution of scientific ideas. The answer favored by psychologists or sociologists, such as Campbell (1960, 1974), is populations of *scientists*. That is, what evolve are not theories but scientific communities. On this view, then, theories as such do not change; rather, scientists change the theories they espouse. The apparent evolution of theories is an artifact.[15]

Of these two approaches I would urge pursuit of the latter. In general, the former view seems a holdover of the old intellectualist, theory-centric orientation that characterized logical empiricist philosophy of science. On the latter, agent-centered view, theories do not have a life of their own. Nor do they exist in a timeless Platonic heaven or in Popper's third world. They must be embodied in people and their artifacts, both abstract and material.

Beyond such generalities, there are more specific reasons for taking scientists, not theories, to be the basic individuals in a theory of science. One is that a scientific theory of science should itself be a *causal* theory. Concepts themselves do not have any causal powers; they cannot *do* anything. But people can. Of course having a particular concept, or believing a particular theory to be correct, is often a causal factor in a scientist's decision to choose one course of action rather than another. But in this ideas are no different than physical skills, say in operating some experimental apparatus, or personality traits, like ambition or risk-aversion. These traits also cannot operate on their own, but only as part of a functioning human being.

The model of biological evolution itself provides another reason for making scientists the basic individuals of an evolutionary account of science. To take a standard example, we say that the neck of the giraffe evolved from shorter to longer. But that way of talking is recognized to be mere shorthand. Individual necks did not get longer. Rather, interaction with the environment made it possible for giraffes with longer necks to have more successful offspring. So the average length of giraffes' necks increased from generation to generation. The operative causal interactions, however, were between individual giraffes and their environment, and with each other. Having a longer neck was just one causal factor, among many, that made some giraffes reproductively fitter than others.

It sometimes happens, as with Marie and Pierre Curie, that literal reproduction between scientists yields new scientists. But even in these rare cases, the child, like other pupils, had to be taught the parents' views. Being a holder of a particular theory is always an *acquired* trait. This is one important respect in which the analogy between biological evolution and the evolution of scientific fields fails to be complete. Populations of scientists sharing a theory grow largely through teaching, indoctrination, and professionalization. They also grow (or decline) through the in (or out) migration of people who, apart from their original training, individually choose whether or not to accept the theory in question.

On a traditional Darwinian account of organic evolution, variation is supplied primarily by mutations which are random relative to the potential contribution to fitness of the modified trait. That is, being potentially beneficial to the organism does not make a mutation any more likely to occur. New theories, on the other hand, are consciously designed to solve particular problems. That theories do often succeed in solving known problems is thus more than a matter of chance. This is true, but there is still much chance involved.

Problems, of course, logically underdetermine the theories that might be developed to solve them. But more than this, the nature of solutions devised by scientists depends to a great extent on the *resources* available to them, both cognitive and material. Cognitive resources include the kinds of models with which the scientist is already familiar through training and experience. Material resources include institutional support and the availability of knowledgeable colleagues. In the typical case, these resources were not brought together in order to solve the problem at hand. They existed for other purposes which might or might not make them useful in devising a viable solution. And even if the problem itself was selected because it seemed amenable to solution using the available resources, the solution is still not uniquely determined. It remains to some nonnegligible degree a matter of chance whether a proposed solution turns out to be viable or not.

It is often remarked that contemporary science is done primarily in small research groups—not by individuals working alone. But this creates no difficulties for an evolutionary model of science. Modern evolutionary theories allow the possibility of *group* as well as individual selection. It is always an empirical matter how much selection is operating on individuals and how much on the groups to which they belong.[16]

Finally, there is the fact that originally inspired Kuhn. Change in science is sometimes quite rapid and dramatic—nothing like the slow, gradual process described by Darwin. Here again recent work in evolutionary theory provides a ready answer. Rather than being gradual, organic evolution is now widely thought to occur in relatively short sprints which "punctuate" long periods of relative equilibrium.[17] The suggestion that these long stable periods are the evolutionary analogs of Kuhn's periods of normal science is both obvious and irresistible. One can have rapid evolutionary change without "revolutions."

Much more must be done to expand the above remarks into a full-fledged evolutionary model of science. And even more work will be required to show that such a model can genuinely explain major episodes in the history of science. Even in its present embryonic form, however, it seems to me a more promising model of scientific change than stage theories, political models, or models of rational progress.

## 6. Cognitive Mechanisms

The fundamental processes in organic evolution are (i) variation, (ii) selection, and (iii) transmission. There is naturally existing *varia-*

*tion* in the traits possessed by individual members of a population. The fitness of individuals to leave offspring varies with variation in traits. The result is *selection* favoring those whose traits make them fitter. The relative prevalence of the favorable trait increases in the population because that trait is disproportionately *transmitted* to the next generation.

One can go far in explaining many facts about the evolution of particular species in terms of these basic processes, as Darwin first showed. Nevertheless, although by the end of the nineteenth century many people were convinced that evolution had taken place, Darwin's own account of the evolutionary process was in low repute. This was because the *biological mechanisms* underlying variation and transmission were poorly understood. The suggestions of Darwin and his followers seemed to many clearly insufficient to the task. What changed in the first half of the twentieth century is that Mendelian genetics provided a promising account of the required mechanisms. The genetical theory of natural selection succeeded where Darwin's own account of natural selection had failed.

The lesson is that for an evolutionary account of scientific progress to succeed, it must be supplemented with an adequate account of the mechanisms underlying the analogous processes of variation, selection, and transmission. There are, I think, two sorts of mechanisms. First are the biological and psychological mechanisms underlying the cognitive capacities of individual scientists. Second are the social mechanisms operative in the social and institutional environment in which scientists work and interact. For a theory of the cognitive capacities of individuals we should look to the cognitive sciences. And for a theory of the social environment we should look to the social sciences, particularly the sociology of science. However, just as ecology is inherently more complex than genetics, so the social sciences are inherently more complex than the cognitive sciences. I will focus on cognitive mechanisms.

The first thing one must realize is that the designation "cognitive science" includes a diverse set of disciplines ranging all the way from cognitive neurobiology to cognitive anthropology. In between are cognitive physiology, cognitive psychology, linguistics, artificial intelligence, and cognitive sociology. For the job at hand, I am most attracted to the "lower" end of this spectrum, neurobiology, physiology, and psychology. Much research in artificial intelligence, in particular, seems to me too committed to a computational picture of thinking that often looks like a computerized version of Logical Empiricism. Humans just do not work like that.[18]

*Representation*

What distinguishes the contemporary cognitive sciences from earlier work on similar topics is a focus on how humans construct, manipulate, and store *representations*.[19] If we are to use this resource in attempting to understand how scientists construct, use, and learn *theories,* we must recast our thinking in appropriate ways. This reorientation makes much of the philosophical literature on the nature of theories obsolete. Most of that literature assumes that theories are to be analyzed as formal, axiomatic systems, which are a particular type of *linguistic* entity.

A standard view in cognitive psychology suggests associating theories with *families of models,* or "schemata." Particular models can, of course, be described linguistically, but there is no reason to take the models themselves to be linguistic entities. They are, rather, more abstract structures which may, however, be physically encoded in networks of neurons. In fact, most of what Kuhn says about "exemplars" holds for schemata, except that here the notion is applied to any model, not merely to those few that play a particularly central role in shaping a scientific field.

On this approach, one gives up the idea of a "theory" as an ontologically definite, well-defined entity. The closest one can come to a definitive presentation of a theory is in the standard, advanced textbooks. These contain most of the exemplars of the field together with a range of other models, but none could claim to present "all" the models. That is not a well-defined notion. Nevertheless, this approach does provide the right kind of underpinning for an evolutionary account of scientific change.[20]

Even in relatively specialized fields, different scientists possess somewhat different repertoires of models which they are prepared to deploy in new situations. These differences may be the result of having gone to different schools, having had different teachers, having had different interests, and having worked on different problems since completing their education. So here we have a ready source of *variation* in the acquired cognitive resources of individual scientists. This variation is surely great enough to provide a considerable part of the explanation why different scientists propose different solutions to the same problem.

*Scientific Judgment*

No one denies that scientists exercise their individual judgment as to which theory or general approach to the field is best. But differences among rival accounts of science are perhaps most strongly revealed when it comes to accounting for how such judgments are made. Logical empiricists, of course, denied any official interest in how such judgments actually are made. They claimed to be uncovering a *logical* relationship between data and theories which objectively determined which of rival theories is best supported. To make a mistake in judgment on such matters is like committing a logical fallacy, such as affirming the consequent. Later philosophers of science, though officially more concerned with judgments actually made, nevertheless focused their energies on developing an objective criterion for the *rational* choice of a research tradition. Kuhn and recent sociologists of science portray such judgments as essentially *political* in nature, the result of negotiations and struggles among competing interests.

Research by cognitive psychologists into human judgment concentrates on how people actually make judgments about various subjects, including causal relationships and probabilities. But psychologists are also interested the *effectiveness* of people's judgmental strategies in leading them to make *correct* judgments. Similarly, they investigate how people could be more effective and why they often are not as effective as they might be given the evidence at their disposal.[21]

This research is sometimes described as an investigation into human *rationality*. But it is rationality in the purely *instrumental* sense of employing effective means (judgmental strategies) to achieve desired goals (correct judgments). The rationality most philosophers have sought has been *categorical* in the sense that whether a judgment is rational is completely independent of whether it turns out to be correct or not. In this sense, rationality is like logical validity. Whether an argument form is valid or not does not depend on the truth or falsity of either the premises or the conclusion. The nature of the relationship is all that matters.

From the standpoint of an evolutionary model of science, the judgments of scientists about particular models, or whole programs of research, constitute one important mechanism which scientists use to select themselves into or out of subpopulations sharing a particular viewpoint. The apparent "evolution" of the corresponding family of models is an artifact. What really happens is that the relative number of scientists in the field advocating models of a particular type increases as the result of the decisions of individual scientists to join this group. Another source of change, of course, is the training of new recruits.

*Experimentation*

Everyone agrees that experimentation has played a major role in the evolution of science since the seventeenth century. Once again, however, different accounts of the nature of science offer very different pictures of the nature and importance of that role.

Classical logical empiricist writings, for example, contain almost no references to experimentation as such. The reason seems to be that, within this framework, the only point to experimentation could be to produce new *data* which could then be used as premises in one's inductive logic. But whether the data came from experimentation or just observation would be irrelevant. The most one could say, as some more recent authors have pointed out, is that one could use an inductive logic to determine what data, if it were available, would strongly confirm or disconfirm various hypotheses. That could provide a motivation for designing experiments which, assuming the best confirmed of existing hypotheses, are most likely to yield the desirable data.

The writings of Kuhn, and of the major post-positivist philosophers of science, exhibit a similar "neglect of experiment."[22] This can only be ascribed to the general "theory centrism" of philosophers and intellectual historians. The same cannot be said of relativist sociologists of science, whose writings are full of detailed accounts of episodes in the laboratory. Almost without exception, however, the point of these investigations is to exhibit "contingencies" in the design, execution, and interpretation of experiments. The point, of course, is to show that the results of experiments do not provide an "objective" or unequivocal basis for choosing among alternative theories. Rather, like theories, experimental results are also subject to negotiations among scientists with competing interests.

Thinking of individual scientific judgment as a selection mechanism provides another account of experimentation. On this account, experimentation provides the individual scientist with a way of choosing among alternative models—a way that gives the world itself a major role, though of course not a totally determining role, in the choice. This way of looking at experimentation focuses attention on the *design* of an experiment as a vital component in a scientist's *decision strategy*. The overall goal of the strategy is not merely to produce data, but to produce data that will sharply *discriminate* among rival models.[23]

## 7. Toward a Naturalistic, Representational, Cognitive, Evolutionary Theory of Science

By way of summary, I offer the following review of the major features I think future theories of science should exhibit.

### *Naturalistic*

A theory of science must itself be a *scientific* theory. Attempts to construct accounts of science *a priori* by appeal to logic or someone's intuitions cannot succeed. There just is no place to ground a philosophically autonomous theory of science. This does not mean that one is reduced merely to describing the behavior of scientists. The goal is to be able to *explain* how science is done and why it is as successful as it is. Nor is one prohibited from making *normative* claims about how to pursue scientific goals effectively. But any such normative claims must be based on reliable empirical theories of how science in fact works, theories that connect methodological strategies with clearly articulated goals.

*Representational*

The models scientists construct are clearly intended by them to *represent* various aspects of the world. And it is very difficult to deny that at least sometimes they are moderately successful at this task. An adequate theory of science must make sense of this fundamental aspect of science. That Kuhn and later historically inclined philosophers of science have been satisfied with a *non*representational account of science seems in large measure an overreaction to the obvious shortcomings of Logical Empiricism. But it is possible to have a representational account without embracing anything like Logical Empiricism.

*Cognitive*

The ability to construct models of complex and often remote aspects of the world is a deliberate and self-conscious extension of the evolved cognitive capacities for "mapping" their environment which humans share with many animals, particularly other mammals. In constructing their own models of science, philosophers of science should exploit this fundamental insight of contemporary cognitive science. The cognitive sciences also provide resources for constructing a naturalistic account of scientific judgment which replaces the inductive logics of Logical Empiricism.

*Evolutionary*

Science does not develop in a series of "stages" in any but a trivial sense. Nor does it exhibit "rational progress" in the sense of Lakatos or Laudan. The growth of science is more *evolutionary,* with individual scientists, together with their ideas, being selected from a population exhibiting considerable variation. The focus should be on the evolution of scientific *communities,* at the level of research specialties, rather than on the evolution of "ideas." The ideas follow the scientists, not the other way around. So the family of models associated with a research specialty evolves as the associated community evolves.

*Conclusion*

If the philosophy of science is itself a theoretical part of a science of science, then, on the above picture, it too should evolve both as a research community and, consequently, as a family of models. I have sketched its evolution since World War II and indicated the ways I would like to see it evolve in the future. Whether it does or not depends, of course, on the individual judgments of people in the field and, perhaps even more so, on the people who will be coming into the field in the future.

# THREE

# Science and Technology Studies

## 1. Introduction

I begin with two curious facts about the whole area of Science and Technology Studies. First, science and technology have been major forces shaping societies for several centuries, particularly during the twentieth century, and particularly since World War II. Yet Science and Technology Studies has only during the past several decades become recognized as a distinct area of study. Second, it is technology more than science that directly influences how people live. Yet within Science and Technology Studies, Science Studies reached a high level of development before Technology Studies received much attention at all. Today the History, Philosophy, and Sociology of *Science* are all still much better developed than the History, Philosophy, and Sociology of *Technology*.

Whatever their causes, one consequence of these historical developments has been that many recent theoretical approaches to Technology Studies take the form of attempts to redeploy in the study of technology models developed earlier in Science Studies.[1] Hardly anyone now thinks of technology as merely applied science, but, in practice, many people treat Technology Studies as applied Science Studies. The assumption behind this strategy is that there are sufficient similarities between science and technology that models developed in Science Studies may usefully be applied in Technology Studies.

Granting a parallel between Science Studies and Technology Studies has some interesting consequences for theoretical debates within Science Studies itself. In particular, it counts against both of the current extreme views in favor of some more moderate view. I will label the extreme views "enlightenment rationalism" and "postmodernism" respectively. This motivates the somewhat tongue-in-cheek label, "enlightened postmodernism," for a projected moderate synthesis. In what follows I will motivate these labels and say why a concern with technology counts against the extreme views and thus for an enlightened postmodern synthesis.

## 2. Enlightenment Rationalism

Among the central tenets of the position I am calling "enlightenment rationalism" are (1) that the material world is governed by universal natural laws, and (2) that the truth of these laws can be known by human agents employing universal principles of rationality. This characterization includes both "rationalists" such as Leibniz and "empiricists" such as Locke. They differ primarily in their understanding of the particular nature of laws and of the required rational principles. What is important here is the implication that science is *autonomous* from the rest of society in that its legitimacy and authority are grounded in universal principles that transcend any particular social context.[2]

The enlightenment rationalist picture of science was inspired by the success of Newtonian physics. But it was also motivated by various political circumstances. In England it was seen as supporting both the restored monarchy and the Church of England. In France the political metaphors in the phrase "governed by laws of nature" served as a powerful rhetorical resource for opposition to absolute monarchy and in favor of "the rule of law."[3]

Logical Empiricism in the philosophy of science may be seen as a twentieth-century embodiment of enlightenment rationalism. It was motivated, in part, by developments in logic and the foundations of mathematics initiated by Russell and Whitehead. But Logical Empiricism was also inspired by the Einsteinian revolution in physics and by the breakdown of European society before and after the First World War (Janik and Toulmin 1973).

Similarly, twentieth-century functionalist sociology of science presupposed an enlightenment rationalist picture of science by limiting the sociology of science to the study of social structures conducive to the practice of science conceived as an enlightenment rationalist enterprise. Merton's early writings also clearly portray science as supporting liberal democratic institutions and opposed to totalitarian regimes.[4] One can also see the enlightenment rationalist picture at work in the "internalist" conception of the history of science. In each field, science was seen as autonomous from society— although perhaps also providing external legitimization for some forms of social organization over others.

If one insists that models of science must apply also to technology, then the inadequacies of the enlightenment rationalist picture of science are becoming ever clearer. The vast majority of students of technology now see technology as intertwined with, and shaped by, the surrounding culture. Instead of universal laws and rationality, one finds in technology time-bound techniques and artifacts, together with the political clash of social and economic interests.[5] If one were simply to apply an enlightenment rationalist picture of science to the study of technology, one would miss most of what are now regarded as salient features of modern technology.

## 3. Postmodernism

Much current work in Science Studies proceeds, explicitly or implicitly, in opposition to an enlightenment rationalist view of science. This is particularly true of those who march under the banner of *social constructivism*.[6] Indeed, to some extent social constructivism can be defined as the negation of enlightenment rationalism. (1) Regarding supposed laws of nature, the social constructivist view would be either (a) that the very idea of a law of nature does not make sense, (b) the idea makes sense, but there simply are no such things, or (c) whether or not there are laws of nature makes no difference to understanding how science works. (2) Regarding the supposed rationality of science, the view would be that there are no privileged principles of scientific rationality—only the natural intellectual abilities which scientists share with everyone else, including stockbrokers and politicians.

Among the most consistently radical of the social constructivists is Steve Woolgar, who explicitly attacks what he calls "the ideology of representation," the idea that the claims scientists make are representations of an independently existing world. What is required, he argues, is an "inversion" of this ideology according to which what are typically taken to be independently existing objects are in fact constituted by the claims—together with the whole social context in which such claims are produced (Woolgar 1988).

Given standard historical periodizations, it is appropriate to classify enlightenment rationalism under the general heading of "modernism." The radical inversion of enlightenment rationalism which pursues a "critique" of science by "deconstructing" the claims of scientists merits the fashionable label "postmodernism."

There has, of course, always been some opposition to the enlightenment rationalist picture of science, most notably the various "romantic" movements of the nineteenth century. And there was a reaction against science and technology following each of this century's World Wars. Most relevant for those of us in Science and Technology Studies right now, however, is the reaction against the enlightenment rationalist picture that followed the post–World War II colonial wars—particularly those in Algeria and Indochina. For many scholars of the Vietnam War generation, science and technology had become weapons in the hands of aggressive and repressive national states. This vision both inspired and sustains the social constructivist critique of science and technology. Whereas enlightenment thinkers saw in Newton's Laws a model for the laws of a just society, social constructivists see scientific laws as every bit as contingent human creations as the laws of society, just or not. And this provides a basis for refusing to grant science a privileged epistemological position.

However, shifting one's focus from science to technology also undermines the radical social constructivist position. The original targets of the social constructivist critique of science were highly theoretical claims about such things as gravity waves (Collins 1985), neutrinos (Pinch 1986), pulsars (Woolgar 1981), and the weak neutral current (Pickering 1984). Here the connection between the postulated entities and experimental reality is indeed somewhat tenuous. But this is not true of typical technological innovations.

Consider the computers which no doubt sit on the desks of most social constructivists. In designing and manufacturing computers, scientists and engineers employ a variety of representations for what is going on inside computers. These include modern versions of old-fashioned circuit diagrams—literally maps of electrical pathways. The success of these endeavors, and the reliability of the product, prove beyond any reasonable doubt that there has got to be something right about these representations. We may not yet know how best to characterize what constitutes successful representation in this context. Certainly no one is going to argue that the mind mirrors nature by being imprinted with the essence of perceived objects. But questioning the very existence of successful representation, however it is ultimately to be understood, is sheer postmodern madness.[7]

## 4. Intermediate Viewpoints

I will now survey three intermediate viewpoints that might, singly or in combination, provide a suitable enlightened postmodern synthesis.

*Naturalistic Realism*

Although many philosophers of science still support enlightenment rationalism, many do not. I will describe a version of one revisionist perspective, *naturalistic realism*.[8] This view downplays the idea that there might be universal natural laws encoded in true general statements. Rather, scientists are seen as engaged in constructing models of the world that apply more or less well to narrower or broader classes of natural systems. Second, naturalistic realism denies that there are universal principles of rationality which could sanction belief in the correctness of any particular model. Rather, scientists employ reasoning and decision strategies common to other activities such as business or politics. But science remains a *representational* activity in that scientists are seen as being more or less successful at constructing models that in fact represent various aspects of the world. Significantly for the present discussion, by rejecting the universalistic pretensions of enlightenment rationalism, naturalistic realism is committed to rejecting the autonomy of science, and is therefore open to extensions into the realm of technology.

*Interest Theory*

As I understand them, the original interest theorists of the Edinburgh School did not deny that science is properly conceived as a representational activity. They only denied that the choice of the best representation involves any special rational principles. Rather, the choices scientists make are determined primarily by their "interests," which might be social, political, professional, or personal.[9]

Although it might be thought contentious, I would argue that most of the virtues of interest theories of science can be incorporated into the framework of naturalistic realism.[10] The main disagreement would simply be over the extent to which decisions about the fit of models to the world are influenced by interests rather than other factors, such as the outcomes of experiments. My view is that this balance varies from case to case and the actual influences can only be ascertained by examining the particular scientific episodes in considerable detail. In some cases experimental data may strongly influence theory choice; in other cases political commitments or professional interests might be dominant. Most cases are mixed.

*Actor-Network Models*

Bruno Latour (1987, 1989) has been the primary spokesman for an actor-network model of technoscience. Here I only wish to raise the question of just how much disagreement there is between Latour and radical constructivists such as his one-time coauthor Woolgar. On the surface there appears to be considerable difference in that Latour has no interest in the wholesale deconstruction of modern technoscience. He is perfectly willing to engage in "realist" discourse about unproblematic entities like microbes and DNA. But if one asks how such entities came to be unproblematic, similarities with more radical versions of social constructivism reappear.

For Latour, microbes or DNA are "actants" which have been enrolled by scientists such as Pasteur or Watson as part of a test of strength against their scientific antagonists. So scientists do "represent" entities, but in the way legislators represent their constituents, not the way an artist represents a landscape. The entities are not "discovered." They are constructed through the process by which scientists build networks of allies to defeat their scientific rivals. Thus, where Latour differs from other constructivists is primarily in his particular theory of how the construction of scientific entities is performed. In place of talk about "social interaction," he provides a lively account of how networks of actants are recruited to achieve victory over rival networks. Once victory is achieved, the entities associated with the victorious side are reified to the extent that they may be treated as "black boxes" in later, related inquiries. These black boxes can always be reopened, but not easily by philosophical or sociological analysis. Only other scientists are likely to have the resources necessary to build networks of allies that could render received entities once again problematical.

## 5. Prospects for an Enlightened Postmodern Synthesis

Latour delights in breaking down standard dichotomies, beginning with that between science and technology.[11] There is, he claims, only one "technoscience." He also erases the standard epistemological distinction between the human scientist as one who represents (pictures) objects and the nonhuman objects that are represented (pictured). Both sorts of entities are equally "actors" in his networks. Thus, like Woolgar, he rejects the "ideology of representation." And this rejection seems to cover even naturalized accounts of representation as found, for example, in the cognitive sciences. Although he rejects the label, Latour's position must be judged as more postmodern than enlightened.

Latour's networks are composed of what are usually regarded as heterogeneous elements. Yet his analysis is *sociologically homogeneous* in that all the analytical categories, such as "actor" or "enrollment," are sociological categories. One can understand the desire of sociologists for a theory of science and technology that is totally sociological. But the subject simply will not allow it. A thoroughgoing sociological analysis, no matter how clever, cannot be a complete analysis of any technological system. And this point can be generalized: No *monotheoretical* account of technological development can be an adequate account.

Much of what rings true in Latour's accounts of scientific and technological controversies can be captured by something like Thomas Hughes's "systems" account of technological development.[12] The idea of a system is still fundamentally a "modernist" notion. Although insisting that all components of a system interact, and must be treated as such, a systems approach nevertheless distinguishes the different components and permits the internal analysis of individual parts, and of particular interactions among selected components of the whole system. Latour's actor-network model abolishes all such distinctions.

I conclude that a combination of naturalistic realism, interest theory, and a systems approach provides at least the beginnings of an enlightened postmodern synthesis. This suggestion presupposes that the ideal treatment of any scientific or technological development is not one that analyzes it from a single, unified perspective, but one that successfully integrates diverse component perspectives.

For a technological system, these component perspectives would typically include at least the following: (1) An understanding of a technological artifact in its scientific-technological context. (2) An understanding of relevant psychological or cognitive features of various actors—inventors, entrepreneurs, managers, consumers, etc. (3) An understanding of relevant *micro*social interactions. (4) An understanding of various *macro*social interactions, including cultural and economic interactions. So the goal is not "unification" within a single perspective, but the "integration" of several different perspectives. Rather than *inter*disciplinary research, one should think in terms of *multi*disciplinary research. But this probably still requires more collaborative research than is the current norm in Science and Technology Studies.[13]

## 6. The Social Shaping of Technology

If we are to give up talking about "the social construction of technology," we need another metaphor that will do the appropriate enlightened postmodern job. I suggest we appropriate the phrase, *the social shaping of technology*, which has been used by both Hughes and MacKenzie. What makes this phrase appropriate, from my point of view, is that it implies that there is *something* to be shaped. Technology is not *just* a social construct, as many social institutions are. It is a complex system in which humans construct artifacts in a context shaped both by the interests of humans and by the underlying physical nature of their artifacts.

## 7. Technology Policy

Everyone agrees that results in Science Studies should have implications for science policy. The case for the policy relevance of Technology Studies is even stronger, since the need for technology policy is more obvious. Thinking about the policy implications of Science and Technology Studies provides a further contrast between a simple postmodern and an enlightened postmodern approach.

Social constructivists have had precious little to say about science or technology *policy*. I suspect this silence is not an accident. If one takes seriously the position that science and technology are social constructs, the only policy advice one can give is to improve one's use of the rhetoric of science and technology to persuade others of one's point of view and to build cohesive social networks. But politicians and policy makers do not need anyone to tell them how to do that. That is already how they operate in any sphere of activity. I suspect that social constructivists are not unhappy with this conclusion. They do not want to create anything that might give government policy makers any more power than they already have.

My enlightened postmodern synthesis, on the other hand, retains a more traditional, modernist relationship between knowledge and policy. Knowledge of how an artifact, such as an electric generator, can be shaped into a technological system, such as an electrical power system, can be useful in shaping other technological systems. Of course there is always the chance that such knowledge might be put to purposes we find undesirable. But that is a problem with many kinds of knowledge. The Enlightenment gamble has always been that the chance of good results is worth the risk of bad results. It is at least arguable that this gamble is now too risky. If so, that point is only obscured by the theory that technological systems are simply social constructs rather than systems of artifacts shaped by human interests.

## 8. Conclusion

My conclusions are conditional. I have assumed that there should be a strong overlap between approaches to Science Studies and approaches to Technology Studies. If we then deny the autonomy of technology from society, we must reject the enlightenment rationalist account of science. Similarly, if we grant that the success of modern technologies rests to some extent on successful representations of nature, then we must reject a social constructivist account of science (as well as of technology). The alternatives are to reject the antecedents of these conditionals or to seek a moderate middle ground.

# PART TWO

Perspectives on Science

# FOUR

# Naturalism and Realism

## 1. Methodological Naturalism

To begin a discussion of naturalism, one can hardly do better than quote a sentence from the later writings of the foremost American champion of naturalism in this century, John Dewey.

> Naturalism [Dewey wrote in 1939] is opposed to idealistic spiritualism, but it is also opposed to super-naturalism and to that mitigated version of the latter that appeals to transcendent *a priori* principles placed in a realm above Nature and beyond experience. (Schilpp and Hahn 1939, 580)

This passage is typical of statements of naturalism in emphasizing what naturalism opposes over what it proposes. But even in the context of a negative characterization, what is the *form* of the opposition? Is it simply a matter of *criticizing* perceived non-naturalistic appeals as being unjustified, incoherent, vacuous, and so on? Or, does a naturalist go on to advance the general thesis that there are no supernatural forces operative in the universe? If so, what could be the basis for such a claim? Could such a claim be substantiated without begging the question against a supernaturalist? Even worse problems arise for the opposition to *a priori* principles if this opposition goes beyond criticism of particular claims to *a priori* knowledge. What sort of argument could one offer for the general conclusion that there are no justifiable *a priori* principles? Clearly any attempt by a would-be naturalist to construct an *a priori* argument that there are no justifiable *a priori* principles would be self-defeating. Even in its simplest forms, therefore, naturalism, as a positive program, seems to have great difficulties just getting started. Nevertheless, while endorsing the critical function of naturalism, I, like most self-described naturalists, would like also to have a positive program.[1]

I propose the label *metaphysical naturalism* for any naturalism characterized in terms of theses about the world supported by *a priori* arguments. It seems to me that the project of defending *metaphysical naturalism* is sufficiently unpromising that I would rather seek another form of naturalism. The form a naturalism I will defend, for which I propose the label *methodological naturalism,* characterizes naturalism not in terms of theses about the world but in terms of a set of *strategies* to be employed in seeking to understand the world. The sorts of arguments required to defend a program are very different from those required to defend theses—the temptation to engage in question begging or self-defeating *a priori* arguments is much mitigated. It remains to be seen if the resulting naturalism is naturalism enough.[2]

Naturalism is a general program for all of philosophy, indeed, for all of life. Here I will focus my concerns on some issues in the philosophy of science: scientific epistemology, the historical development of science, and scientific realism. I will begin by illustrating several naturalistic strategies. These provide a basis for further characterizing and defending a naturalistic program for the philosophy of science. I will then go on to propose a resolution to one apparent conflict that arises between naturalism and a realistic understanding of the theoretical sciences.

## 2. Naturalistic Priority

One of my proposed methodological strategies for naturalism is what I shall call *Naturalistic Priority*. It is best illustrated by the historically most important exemplar for any form of naturalism, Darwin's explanation of the origin of species. The historical background to Darwin's theory included a strong tradition of natural theology in which the design of nature, particularly the design of animals and humans, was taken as evidence for a supernatural designer and creator. Darwin, as I understand him, did not provide direct evidence *against* the existence of such a creator. Rather, his theory of evolution by natural selection provided an *alternative,* naturalistic explanation of the acknowledged facts, for example, of the fit between the functional anatomy of animals and their environment. The arguments of natural theology were thereby undercut. The existing adaptation of organisms to their environments could be explained without recourse to a divine designer. Ordinary causal mechanisms producing variation, selection, and transmission of traits could do the job.

This line of argument was well-known in the American philosophical community by the beginning of the twentieth century. Here, for example, is William James's version in his famous lectures on pragmatism, first published in 1907.

> God's existence has from time immemorial been held to be proved by certain natural facts. Many facts appear as if expressly designed in view of one another. Thus the woodpecker's bill, tongue, feet, tail, etc., fit him wondrously for a world of trees, with grubs hid in their bark to feed upon. . . . It is strange, considering how unanimously our ancestors felt the force of this argument, to see how little it counts for since the triumph of the Darwinian theory. Darwin opened our minds to the power of chance-happenings to bring forth "fit" results if only they have time to add themselves together. He showed the enormous waste of nature in producing results that get destroyed because of their unfitness. He also emphasized the number of adaptations which, if designed, would argue an evil rather than a good designer. *Here,* all depends on the point of view. To a grub under the bark the exquisite fitness of the woodpecker's organism to extract him would certainly argue a diabolical designer. (James 1907, Lec. III)

In the paragraphs following the passages just quoted, James discusses a standard liberal theological response to Darwin based on the supposition that natural selection was the mechanism God employed to produce organic life on earth. Darwin had merely discovered another of God's laws for nature. James dismisses this response with the remark that "this saves the form of the design argument at the expense of its old easy human content."

The strategy of naturalistic priority exemplified in this case might be expressed as follows: *Search* for naturalistic explanations of recognized phenomena, and *prefer* a naturalistic over any supernaturalistic or otherwise non-naturalistic explanation. Left open in this characterization is what constitutes a naturalistic, as opposed to a non-naturalistic, explanation; what makes a naturalistic explanation actually preferable rather than merely possible; why one should search for naturalistic explanations and prefer them when found; and, indeed, how one knows when a naturalistic explanation has been found. The possibility of human evolution had been discussed long before Darwin. Only with Darwin did evolution become a viable option widely regarded as actually undercutting supernaturalism. I will return to these questions shortly. First, another naturalistic methodological strategy.

### 3. The Naturalistic Explanation of Epistemological Norms

Any form of naturalism will rely heavily on both the methods and the results of current science. It is ironical, therefore, that, for many, a primary objection to naturalism rests on the claim that the practice of science itself depends on extrascientific foundations. In particular, it is often thought that, since the methods of science provide a *normative* basis for claims to knowledge, these methods cannot be understood in purely naturalistic terms.[3]

As I see it, whether one can give a naturalistic explanation of the epistemological norms of science depends on the assumed character of those norms. If one presumes that such norms are *categorical* methodological rules, then, indeed, no satisfactory naturalistic explanation is possible. On the other hand, if one understands the norms of scientific epistemology to be *conditional* norms, a naturalistic explanation is possible. The naturalistic strategy is: Explain epistemological norms as statements of instrumentally rational procedures with means appropriate to the achievement of recognized cognitive goals.

Before saying anything more about the epistemological norms of science, I must say something about my conception of the objects to which those norms apply. Much philosophy of science presupposes a framework in which the focus is on linguistic entities, statements, and in which the connection between statements and the world is understood in terms of the notions of reference and truth. For many reasons, which I will not develop here, I think these notions too crude a basis for a satisfactory understanding of science. A richer, more satisfactory, picture results from introducing intermediate representational entities, for which I use the designation "models." Models may be characterized using statements, but these function in this context merely as definitions. Models may also be characterized, often only partially, using *non*linguistic means, such as diagrams or physical scale models.

On this view, the empirical representational relationship is not directly between statements and the world, but between models and the world. Here the operative notion is not truth, but similarity, or "fit," between a model and the world. Of course one can formulate the hypothesis that the model fits the world and ask whether this hypothesis is true. But such uses of the concept of truth can be understood in a purely semantic, redundant fashion. To say it is true that the model fits is merely a metalinguistic way of saying that the model fits. The former phrase adds no content not already contained in the latter.

Returning to the problem of providing a naturalistic explanation for epistemological norms, let us again begin with reference to a historical case, the discovery of the double helix. According to a standard narrative, Watson proposed a three-chain model of DNA and built a working model which he and Crick showed to Rosalind Franklin and her colleagues at King's College, London. Franklin immediately pointed out that this model could not possibly represent DNA. Measurements of the water content of DNA samples showed ten times the amount of water that could be accommodated by Watson's model. A year later, Watson and Crick came back with a two-chain model that provided appropriate places for water molecules to attach themselves. More importantly, calculations based on the two-chain model showed that a molecule with this structure would diffract X rays in a very distinctive pattern—a pattern Franklin had already found experimentally. All sides agreed that DNA must indeed have a two-chain structure similar to that exhibited in Watson and Crick's latest model. As Watson and Crick put it, they had discovered "the secret of life."

I would not deny that there are ways of reconstructing the reasoning in this narrative that appeal to *a priori*, or otherwise extra-scientific, principles. The task for a naturalist is to produce a reconstruction that avoids any such appeal. This I will now do following the strategy stated above. The implicit goal in this narrative is the construction of a model that can be shown to exhibit a good fit to the actual structure of DNA. This goal breaks down into two coordinate goals: the rejection of models that do not fit, and the adoption of models that do fit.

Consider the reasoning that led to the rejection of Watson's three-chain model of DNA. Franklin's experiments had revealed the relative water content of DNA samples. That constitutes the *data* in this case. The model itself determines what the relative water content of DNA would be if its structure were like that of the model in the relevant respects. Following traditional terminology, I call this a *prediction* from the model. The epistemological decision rule is simple. Since the data and the prediction disagree, decide that the model fails to fit the actual structure of DNA. The reasoning here concerns only the relationship between means and ends. If Watson had achieved his goal of constructing a well-fitting model, the data and prediction should have agreed. But they did not. So apparently he did not achieve his goal. This form of reasoning is effective in that it promotes the goal of not adopting models that do not fit. One may point out that this summary has the form of *modus tollens*. And indeed it does. But that realization, I think, adds little to our understanding of the reasoning in this episode.

The positive case is more complex. The two-chain model indeed yields the prediction that DNA will diffract X rays in the very specific way Franklin had already observed. The decision strategy of using the observed agreement between data and prediction to decide that the two-chain model provides a good fit to the structure of real DNA molecules, however, requires *more* than just such agreement. If many different, but initially equally plausible, models would yield a similar prediction, following such a decision strategy could all too easily lead one to conclude that a model fits when it does not. That is why mere agreement with the measured water content was not regarded as evidence for the two-chain model. Many different models can satisfy that demand. What saved the day in this case was the prior judgment that there are no plausible alternative models that predict the highly specific observed X-ray pattern. That is, features exhibited in the X-ray pattern are quantitatively related to the pitch of the helix in a double helical structure. No other models under consideration could mimic that result. In this case, agreement between data and prediction is a good indicator of fit between model and world, or, more cautiously, a good indicator that the proposed model *fits better* than any others regarded as plausible rivals.

I do not claim that this is the only way of understanding the epistemology of science naturalistically. That there is one way is enough to show that a naturalized epistemology of science is possible. It is important to emphasize, however, that conditional norms for deciding which of two models best fits the world do not require any backing by categorical norms. We do not require a categorical norm such as: Thou shalt decide which model best fits the world, or, more generally, Thou shalt do science. There is no such categorical imperative. One always does science because of some other goal. To put it in a slogan: Scientific rationality is instrumental rationality.

## 4. Naturalism and Pragmatism

It was not just eccentricity or patriotism that led me earlier to quote Dewey and James. I think that there are in fact strong internal connections between naturalism and pragmatism. One of the major themes in pragmatist thought, going back to Peirce, is that it is a mistake to pursue foundationalist programs in epistemology, such as that exemplified by Descartes's method of universal doubt. Rather, we should take our current beliefs as provisionally sound, but only fallibly so. Any particular belief can become the object of motivated doubt. Such doubts initiate inquiries into the grounds for the particular belief in question, inquiries which necessarily take for granted many of our other beliefs. The particular inquiry ends when doubts about the belief under investigation are resolved one way or the other. In sum, our epistemological concerns should focus not on the ultimate grounds for our beliefs but on the methods used to question and revise our beliefs.

Now let us return to the question of what constitutes a *naturalistic* explanation. The paradigm naturalistic explanation is a *scientific* explanation. But what, then, constitutes a scientific explanation? A pragmatic orientation suggests that it would be a mistake to embark on a search for a universal criterion to demarcate science from non-science. Rather, we should provisionally take the recognized sciences of our own time as legitimate. If specific doubts are raised about any of them, these doubts can then be investigated. On this basis, evolution by natural selection has generally been counted as a scientific explanation while special creation has not.

This attitude corresponds well with the historical record which shows that what counts as a scientific explanation changes over time. For most of the seventeenth century, for example, mechanical explanations appealing to action at a distance would have been rejected. In the eighteenth century, after the impact of Newton's *Principia,* such explanations became commonplace. Ultimately, a naturalist can do no more than follow such historical developments, although, of course, the option of questioning any particular historical development always remains open. This leaves methodological naturalism open ended, but not vacuous, as some have implied.[4] At any particular time, some sorts of models will fall within a naturalistic or scientific framework and others will not. Some pursuits will count as naturalistic and others not.

Similar comments apply to my understanding of scientific epistemology. Applying the decision rules outlined above requires substantial knowledge about the experimental situation. In the DNA episode, it had already to be known that Franklin's techniques yielded reliable knowledge of the water content of DNA and of the pitch of a helical molecular structure. If this knowledge were based on previous experiments, they too would have required appropriate prior knowledge. Pursuing this line of argument leads to a quest for the holy grail of a foundational inductive method that can be applied with no prior general knowledge whatsoever and whose use could be justified *a priori*. A pragmatic attitude recommends that this regress be allowed to end where there is no longer any active doubt. No foundation is required.

Finally, pragmatism suggests a naturalistically responsible answer to the question, "Why pursue a naturalistic program?" One may pursue a naturalistic program because of adherence to an ideal that all explanations of workings in the world should be naturalistic in the sense of either being explicitly scientific or else being everyday historical accounts which make no overt appeals to supernatural powers. Now, anyone adhering to such an ideal cannot but harbor beliefs regarding the viability of the program and its eventual success. Moreover, one may regard one's adherence to a naturalistic program as justified by appeal to past successes at finding naturalistic explanations. One might even argue that the success rate has been going up for the past three hundred years. Such beliefs may be said to constitute a *theoretical naturalism,* a doctrinal companion to *methodological naturalism*. A naturalist must be careful, however, not to be tempted into offering *a priori* arguments for the claims of theoretical naturalism. In any case, the real payoff for any theoretical naturalism, as pragmatists would insist, is in its recommendations for further research, which include the strategies of methodological naturalism.

## 5. Naturalism and Realism

My account of scientific epistemology pits one model, or family of models, against rival models, with no presumption that the whole set of models considered exhausts the logical possibilities. This means that what models are taken best to represent the world at any given time depends on what rival models were considered along the way. And this seems, historically, a contingent matter. So the models of the world held at any given time might have been different if historical contingencies had been different. On the other hand, most scientists and philosophers, and not just scientific realists, regard the truths about the world as being fixed. So, what reason could there be for thinking that what we have now are truths if we could now hold different models of the world? Similarly, what reason could we have for thinking that we are, over time, getting closer to the truth? Thus, my pragmatically inspired naturalism seems incompatible with scientific realism.

This problem is not merely an artifact of my particular understanding of scientific epistemology. The idea that science proceeds by comparing non-exhaustive alternatives has been a standard doctrine in the philosophy of science ever since Kuhn (1962) argued that a paradigm is never abandoned in the absence of an alternative. It subsequently became an integral part of the methodological systems of both Lakatos (1970) and Laudan (1977). Significantly, neither Kuhn, nor Lakatos, nor Laudan embraced scientific realism. Indeed, Laudan (1981) has been an outspoken critic of scientific realism. Moreover, pragmatists in the tradition of Dewey have been more identified with instrumentalism than with realism.

My approach to this problem is to admit the contingent nature of scientific development and revise the conception of scientific realism. In fact, the standard conception of scientific realism seems to me largely a carryover from an older picture of science. It is time for this picture to be revised in any case. The clearest expression of the old picture is to be found in the philosophy of mathematics and logic, and in formal semantics. The idea is that the structure of reality mirrors the structure of set theory. Reality is conceived of as consisting of discrete objects, sets of discrete objects, sets of sets of objects, sets of ordered pairs of objects, and so on. True statements are those that describe objects as belonging to the sets to which they in fact belong. A complete science would be the conjunction of all and only the true statements about this set-theoretically structured reality.

This scheme works quite well for understanding purely mathematical entities, where only abstract structure matters and empirical meaningfulness not at all. The assumption among philosophers of science for most of this century has been that the same sort of scheme works for empirical science as well. The tacit argument seems to have been that physics is the fundamental science, and physics is highly mathematical, so our notions of reference and truth developed for mathematics must work also for physics, and, by implication, for all of science. One need only openly state this argument to realize that this picture of science is not necessarily correct.

Here is an alternative picture. Imagine the universe as having a definite structure, but exceedingly complex, so complex that no models humans can devise could ever capture more than limited aspects of the total complexity. Nevertheless, some ways of constructing models of the world do provide resources for capturing some aspects of the world more or less well. Other ways may provide resources for capturing other aspects more or less well. Both ways, however, may capture some aspects of reality and thus be candidates for a realistic understanding of the world. So here, in principle, is a solution to the problem of finding a picture of science that is both naturalistic and realistic. It does not matter that different historical paths might lead to different sciences. Each might genuinely capture some aspects of reality.

This way of presenting the alternative picture makes it seem like a piece of *a priori* metaphysics, something no methodological naturalist can allow. The same picture can, however, be presented as a methodological precept: Proceed as if the world were this way. As such it needs no *a priori* justification. Its vindication, if any, would be in its success. I use the term "perspective" to designate a way of constructing scientific models. Within a perspective, some models may be discovered experimentally to fit the world better than others, at least for some purposes. Comparisons across perspectives are more difficult, but I doubt one ever finds the kind of radical incommensurability that philosophers such as Kuhn and Feyerabend are supposed to have advocated. In any case, the resulting view of realism can be called *perspectival realism*. I will give some examples shortly. My main task now is to make plausible a realistic understanding of science that is perspectival rather than absolute. Here I go well beyond the bounds of traditional pragmatism.

## 6. Perspectivalism in Observation

Perspectivalism is most easily appreciated at the level of observation. It is here that the metaphor of perspectives is closest to its roots. A paradigm example would be the experience of viewing a building from different angles and different distances. Each distance and angle of view provides a different perspective on the building. Two features of this simple case already exemplify the main features of a more general perspectivalism. First, there is no total or universal perspective, or, alternatively, there is no perspective from nowhere or from everywhere at once. All perspectives are partial relative to their objects. Second, each perspective is a perspective of the building. There is something real that each perspective is a perspective of. So perspectivalism is *prima facie* a form of realism, not relativism or constructivism. Additional objectivity can be built into this example by imagining a series of photographs taken from different viewpoints rather than simply a series of visual experiences.

The example of secondary qualities stretches the metaphor of perspectivalism a bit further. We perceive the world as containing objects exhibiting different colors. Yet we know it is our capacity for experiencing color that partly explains the character of our perceptions. Without perceivers like us, there would be no experiences of color. Nevertheless, our perception of color provides us access to aspects of the world apart from us, namely differential reflection of light waves of varying frequency. I would say that our capacity for color vision provides us with a perspective from which we can experience aspects of the world. This perspective is not universal even among humans. Some humans suffer from achromatopsia, a genetic abnormality which results in retinas with only rods and no cones. Such humans have no experience of color whatsoever.[5] It is again a further extension of the metaphor to say that different species of animals experience the world from the different perspectives provided by their various sense organs.

The existence of scientific instrumentation provides a further extension of the metaphor. Radio telescopes, for example, may be said to provide us with a perspective from which to view the heavens. It is a different perspective from that provided by more ordinary optical telescopes. Without this technology, the kinds of outputs provided by such instruments would not exist. Yet radio telescopes do provide us with information about aspects of the universe that may not be accessible in other ways. Similar comments apply to the infrared detectors aboard the Hubble Telescope which recently provided images of the so-called Pistol Star at the center of the Milky Way.

These examples suggest that all forms of observation or detection should be understood as perspectival in nature. They all provide access to reality, access that is, nevertheless, always partial.

## 7. Perspectivalism in Theories

The extension of perspectivalism to the level of scientific theory is more problematic, but, I think, eventually equally convincing. As an intermediate step, consider the human practice of making *maps*.[6] Maps, I would say, represent spatial regions from particular perspectives determined by various human interests. Imagine, for example, four different maps of Manhattan Island: a street map, a subway map, a neighborhood map, and a geological map. Each, I would say, represents the island of Manhattan from a different perspective, appropriate, for example, for a taxi driver, a subway rider, a social worker, and a geologist.

Maps exhibit the two primary characteristics I earlier ascribed to perspectives. First, they are always *partial*. There is no such thing as a complete map. Second, maps may be maps *of something*. So maps can be understood realistically. Unlike perceptual experience or the operation of physical detectors, however, the production of a map is an act of deliberate construction. If a nervous system or other physical detector is functioning properly, it automatically registers the information available to it from its particular perspective. Insuring that a map correctly represents the intended space requires much deliberate care. Mistakes can easily be made. Moreover, one can deliberately construct mistaken maps, or even maps of completely fictional places.

One is tempted to ask: How do maps represent physical spaces? Asking the question this way suggests that the answer is to be found in some binary relationship between maps and places. A better question is: How do we humans manage to use maps to represent physical spaces? This way of posing the question makes it less easy to forget that making maps is a cognitive and social activity of humans.

Part of the answer is that map-making and map-using takes advantage of similarities in spatial structure between features of a map and features of a terrain. But one cannot understand map-making solely in terms of abstract, geometrical relationships. Interpretive relationships are also necessary. One must be able to understand that a particular area on a map is intended to represent, for example, a neighborhood rather than a political division or corporate ownership. These two features of representation using maps, similarity of structure and of interpretation, carry over to an understanding of how humans use scientific models to represent aspects of the world.

It is not too great an analogical leap, I think, to go from maps to the kinds of models one finds in many sciences. Classical mechanics provides an archetypal example. The models of classical mechanics represent only a very limited number of aspects of objects, such as mass, position, velocity, and acceleration. This being so, I would say that the principles of classical mechanics provide a perspective within which one can construct a wide variety of models, some of which we have found to provide a very good fit to mechanical systems in the real world.

Consider, for example, textbook representations of a simple harmonic oscillator. These typically contain numerous equations used to characterize the model. But they also often include a number of diagrams which look a lot like maps. That is, these diagrams are made up of lines in two dimensions which represent various aspects of the motion of a simple harmonic oscillator. One need only know how to interpret them.

Here, then, is a suggestion for better understanding what it means for a model to fit the world. The fit between a model and the world may be thought of like the fit between a map and the region it represents. How further to understand that sort of fit is, unfortunately, not an easy question.

## 8. One World as a Methodological Rule

A final problem. Suppose there are two perspectives that overlap, but disagree about the character of the world in the region where they overlap. What should one say? A pragmatist might try to brush off the question with the reflection that, since we are only concerned with applying models, what does it matter if applications conflict? Use the type of model that best serves one's current purposes. I find that attitude understandable, but too instrumentalistic.

The basis for my misgiving is the supposition that there is, after all, only one world, and it has some one structure or other. Of course, from a naturalistic perspective, one cannot offer transcendental arguments in favor of a "one world" hypothesis. One can, however, take it as a methodological rule: Proceed as if the world has a single structure. In light of this rule, the existence of conflicting applications of different types of models is an indication that one or both types of models fail to fit the world as well as they might. It is an invitation to further inquiry to find models that eliminate the conflict, although there is no guarantee that such models will be found. If arguments in favor of adopting this methodological rule are desired, the best one can do is to point out cases in the history of science where following such a rule was fruitful in leading to the construction of better models.

## 9. Conclusion

I conclude that naturalism is a viable project, not only for the philosophy of science in general, but also for a *realistic* philosophy of science. It is also, I think, a viable project for other areas of philosophy: mind, language, morals; even logic and mathematics. One must be careful, however, how one characterizes naturalism. Presented as theses, it invites self-defeating attempts to construct transcendental arguments in its favor. Vigilance is required to keep arguments for naturalism within naturalistic limitations. A better strategy, I have suggested, is to focus on naturalism as a set of methodological rules for developing a comprehensive naturalistic picture of the world. Success in applying these rules gives comfort to those pursuing the program and encourages others to join the effort. Positively, that is the most a consistent naturalist can do.

# FIVE

# Science without Laws of Nature

## 1. Interpreting the Practice of Science

It is fact about humans that their practices are embedded in interpretive frameworks. This holds both for individuals and for groups engaged in a common enterprise. Of course, any sharp distinction between practice and interpretation, whether drawn by participants or third-party observers, will be somewhat arbitrary. Nevertheless, drawing some such distinction is useful, perhaps even necessary, for those who, while not direct participants in a practice, seek to understand it from their own perspective.

Such is the situation of historians and philosophers of science regarding the practice of science and the concept of a *law of nature*. The claim of some philosophers, for example, that scientists seek to discover laws of nature, cannot be taken as a simple description of scientific practice, but must be recognized as part of an interpretation of that practice. The situation is complicated, of course, by the fact that, since the seventeenth century, scientists have themselves used the expression "law of nature" in characterizing their own practice. The concept is thus also part of an interpretive framework used by participants in the practice of science. That shows that the concept sometimes lives in close proximity to the practice, but not that it is divorced from all interpretive frameworks.

Insisting on the interpretive role of the concept of a law of nature is important for anyone like myself who questions the usefulness of the concept for understanding the practice of contemporary science as a human activity. I realize full well that many others do not share this skeptical stance. Being part of the characterization of the goals of science is but one interpretive role played by this ubiquitous concept. Laws played an essential role in Hempel's (1948, 1965) influential analysis of scientific explanation, and they continue to play a central role in more recent accounts (Salmon 1984). Nagel's (1961) classic analysis of theoretical reduction focuses on the derivation of the laws of one theory from those of another theory. Even critics of these analyses, including radical critics (Feyerabend 1962), have generally focused on other features and left the role of laws unexamined. A concern with the status of laws has inspired many investigations into the confirmation or falsification of universal statements. Laws also figure in contemporary analyses of the concept of determinism (Earman 1986). And scientific realism is often characterized in terms of the truth, or confirmation, of laws referring to theoretical entities.

It is thus not surprising that, like Kant two centuries ago, many contemporary philosophers take it as given that science yields knowledge of laws of nature. Their problem, like Kant's, is to show how such knowledge is possible.[1] For one who doubts that such knowledge is actual, the problem of how it might be possible is not urgent. More serious would be claims that knowledge of laws is not only actual, but necessary for understanding the practice of science. I shall not here be concerned to rebut such arguments.

I will begin by advancing some general reasons for skepticism regarding the role of supposed laws of nature in science. Then I will outline an alternative interpretive framework which provides a way of understanding the practice of science without attributing to that practice the production or use of laws of nature as typically understood by contemporary philosophers of science. Finally, I will sketch explanations of how some expressions can play a fundamental role in science without being regarded as "laws," and how one can even find necessity in nature without there being "laws of nature" behind those necessities. I shall thus be offering an interpretation of science even more radical than what David Armstrong once called the "truly eccentric view . . . that, although there are regularities in the world, there are no laws of nature" (1983, 5). On my interpretation, there are both regularities and necessities in nature, but there are no laws of nature.

## 2. Historical Considerations

One way of understanding the role that a concept plays in an interpretation of a practice is to examine the *history* of how that concept came to play the role it now has. Through the history one can often see the contingencies that led to that concept's coming to play the role it later assumed and realize that it need not have done so.

Of course there is a standard answer to this sort of historical argumentation. The origins of a concept, it is often said, are one thing; its validity quite another. Philosophy is concerned with the validity of a concept, whatever its origins.[2] But this answer rings somewhat hollow in the present context. It is typically assumed that we need a philosophical analysis of the concept of a law of nature *because* that concept plays an essential role in our understanding of science.[3] Inquiring into how the concept came to play its current role may serve to undercut this presupposition.

Among the characteristics attributed to laws of nature by contemporary philosophers of science, several are especially prominent. Laws of nature, it is typically said, are *true* statements of *universal* form. Many would add that the truths expressed by laws are not merely contingent, but, in some appropriate sense, *necessary* as well. Finally, laws are typically held to be *objective* in the sense that their existence is independent of their being known, or even thought of, by human agents.[4]

These characteristics, I believe, came to be associated with some scientific claims not primarily through reflection on the practice of science, but in large part because of the participation of particular individuals in the culture of Europe, mostly in the seventeenth century when modern science began to take the form it now exhibits. The evidence for this claim now seems to me substantial. An extensive review of the relevant literature, however, is beyond the purposes of this chapter. Here I can offer only a brief review and further references.

The main sources for the later use of "laws of nature" as a concept to interpret the practice of science are to be found, it seems, in the works of Descartes and then Newton. For both, the laws of nature were prescriptions laid down by God for the behavior of nature. From this premise the predominant characteristics of laws of nature follow as a matter of course. If these laws are prescriptions issued by God, the creator of the universe, then of course they are true, hold for the whole universe, are necessary in the sense of absolutely obligatory,[5] and independent of the beliefs of humans, who are themselves subject not only to God's laws of nature, but to His moral laws as well.

There is at least one place in Newton's writings where this line of reasoning is explicit. In an unpublished draft of Query 31 of the *Optics*, dating from around 1705, Newton draws on his conception of the deity to support the universality of his laws of motion. "If there be an universal life and all space be the sensorium of a thinking being who by immediate presence perceives all things in it," he wrote, "the laws of motion arising from life or will may be of universal extent."[6] The modesty with which the connection is here asserted was appropriate. What *empirical* evidence did Newton have for the universality of his laws of motion? Only terrestrial motions, such as falling bodies, projectiles, and pendulums, and the motions of the Sun, Moon, planets, and, allowing the investigations of Edmund Halley, perhaps comets as well. The fixed stars posed a definite problem, for what prevented the force of gravity from pulling all the stars together into one place? Newton had need of his God.

Despite some arguments to the contrary, it seems pretty clear that the idea of laws of nature as emanating from the Deity did not originate with Descartes and Newton, or even in the seventeenth century at all. Nor were all earlier uses of such notions necessarily connected with that of a personified lawgiver. The distinction between divine laws for humans as opposed to laws for the rest of animate or inanimate nature can be traced back at least to Roman thinkers. On the other hand, by the thirteenth century, Roger Bacon seems to have thought of the laws of optics, reflection and refraction, in very much the secular way that became commonplace in the nineteenth century. Galileo is famous for his employment of the "two books" metaphor in which God is portrayed as the author of both the Bible and the "Book of Nature." But the idea of "laws of nature" in the sense later used by Descartes and Newton seems to have played only a minor role in his understanding of the new science. Robert Boyle, who shared many of Newton's theological beliefs, nevertheless urged caution in using the notion of laws of nature on the grounds that, strictly speaking, only moral beings, and not inanimate matter, can appreciate the meaning of laws. One finds similar qualms in the writings of Aquinas.[7]

There is another factor in the story which seems relatively distinct from theological influences, namely, mathematics. Would the concept of laws of nature have gained such currency in the absence of simple mathematical relationships which could be taken express such laws? And do not the qualities of universality and necessity also attach to mathematical relationships? These questions are as difficult as they are relevant. Galileo had the mathematical inspiration, but apparently did not think of the book of nature as containing "laws." Kepler, on the other hand, thought of laws in somewhat the same way as Descartes and Newton. Clearly the theological and mathematical influences both push in the same direction. In any case, the one does not exclude the other. Perhaps both were necessary for the notion of a law of nature to have become so prominent in seventeenth century scientific thought.[8]

In the end, one may still ask why Descartes and Newton were so strongly inclined to interpret various mathematical relationships as expressions of God's laws for nature when thinkers a century earlier or a century later were far less inclined to do so. Here I can do no better than appeal to "the characteristic seventeenth-century blend of a voluntarist theology and a mechanistic, corpuscularian physics" (Milton 1998, 699). What matters most for my purposes, however, is not which ideas one can find when. At almost any period in history one can find a vast range of ideas existing simultaneously. The important question is which of the variety of ideas available at an earlier period got adopted and transmitted to later periods, and thus shaped later interpretations. Here there can be no serious doubt that, for Descartes and Newton, the connection between laws of nature and God the creator and lawgiver was explicit. Nor can there be any doubt that it was Newton's conception of science that dominated reflection on the nature of science throughout the eighteenth century, and most of the nineteenth as well.[9]

The secularization of the concept of nature's laws proceeded more slowly in England than on the continent of Europe. By the end of the eighteenth century, after the French Revolution, Laplace could boast that he had no need of the "hypothesis" of God's existence, and Kant had sought to ground the universality and necessity of Newton's laws not in God or nature, but in the constitution of human reason. Comte's positivism found a large audience in France during the middle decades of the nineteenth century. But, in spite of the legacy of Hume, whether the laws of nature might be expressions of divine will was still much debated in the third quarter of the nineteenth century in Britain. Here the issue was whether Darwin's "law of natural selection" might just be God's way of creating species. Not until Darwin's revolution had worked its way through British intellectual life did the laws of nature get effectively separated from God's will.[10]

It is the secularized version of Newton's interpretation of science that has dominated philosophical understanding of science in the twentieth century. Mill and Russell, and later the Logical Empiricists, employed a conception of scientific laws that was totally divorced from its origins in the theological climate of the seventeenth century. The main issue for most of this century and the last has been what to make of the supposed "necessity" of laws. Is it merely an artifact of our psychological makeup, as Hume argued; an objective feature of all rational thought, as Kant argued; or embedded in reality itself?[11]

My position, as outlined above, is that the whole notion of "laws of nature" is very likely an artifact of circumstances obtaining in the seventeenth century. To understand contemporary science we need not a proper analysis of the concept of a law of nature, but a way of understanding the practice of science that does not simply presuppose that such a concept plays any important role whatsoever.

## 3. The Status of Purported Laws of Nature

What is the status of claims that are typically cited as "laws of nature"—Newton's Laws of Motion, the Law of Universal Gravitation, Snell's Law, Ohm's Law, the Second Law of Thermodynamics, the Law of Natural Selection? Close inspection, I think, reveals that they are neither universal nor necessary—they are not even true.[12]

For simplicity, consider the combination of Newton's Laws of Motion plus the Law of Universal Gravitation around the year 1900, before the advent of relativity and the quantum theory. Could one find, for example, any two bodies, anywhere in the universe, whose motions exactly satisfied these laws? The most likely answer is "no." The only possibility of Newton's Laws being precisely exemplified by our two bodies would be either if they were alone in the universe with no other bodies whose gravitational force would affect their motions, or if they existed in a perfectly uniform gravitational field. The former possibility is ruled out by the obvious existence of numerous other bodies in the universe; the latter by inhomogeneities in the distribution of matter in the universe. But there are other reasons as well for doubting the precise applicability of the laws. The bodies would have to be perfectly spherical, otherwise they could wobble. They could have no net charge, else electrostatic forces would come into play. And they would of course have to be in "free space"—no atmosphere of any kind which could produce friction. And so on and on.[13]

Many excuses have been given for not taking more seriously the lesson that, strictly speaking, most purported laws of nature seem clearly to be false. A recent one is that the laws actually discussed by scientists are not the "real" laws of nature, but at best "near" laws.[14] Here I wish only to examine a view that does take the lesson seriously, but remains still too close to the traditional account. This is the view, developed by Coffa (1973) and Hempel (1988), that laws are expressed not by simple universal statements, but by statements including an implicit "proviso." As I understand it, Coffa's and Hempel's account is that purported statements of laws of nature of the form "All bodies, ..., etc." are to be interpreted as really of the form "All bodies, ..., etc., with the proviso that... "My objection to this interpretation is that it is impossible to fill in the proviso so as to make the resulting statement true without rendering it vacuous.

This problem is particularly evident in cases where the implicit proviso must be understood to be expressed in concepts that are not even known at the time the law containing the implicit proviso is first formulated. Most of the laws of mechanics as understood by Newton, for example, would have to be understood as containing the proviso that none of the bodies in question is carrying a net charge while moving in a magnetic field. That is not a proviso that Newton himself could possibly have formulated, but it would have to be understood as being regularly invoked by physicists working a century or more later.[15] I take it to be a *prima facie* principle for interpreting human practices that we do not attribute to participants claims that they could not even have formulated, let alone believed.

It is important to realize that my objection is not just that the proviso account introduces indefiniteness into our interpretation of science. One of the major lessons of post-positivist philosophy of science is that no interpretation of science can make everything explicit. Important aspects of the practice of science must remain implicit. The issue is where, in our interpretation of science, we locate the unavoidable indefiniteness. The proviso account locates indefiniteness right in the formulation of what, on that account, are the most important carriers of the content of science, namely, its laws. I think a more faithful interpretation would locate the indefiniteness more within the practice of science and leave its products, including its public claims to knowledge, relatively more explicit.[16]

## 4. Models and Restricted Generalizations

Let us return to the example of Newton's equations of motion together with the equation for the force of gravity between two bodies. My reference here to Newton's *equations* of motion rather than his *laws* of motion is deliberate. Everyone uses these equations. The issue is how to interpret them, whether as "laws" or as something else. Interpreting the equations as laws assumes that the various terms have empirical meaning and that there is an implicit universal quantifier out front. Then the connection to the world is relatively direct. The resulting statement is assumed to be either true or false.

On my alternative interpretation, the relationship between the equations and the world is *indirect*. We need not initially presume either a universal quantifier or empirical meaning. Rather, the expressions need initially only be given a relatively abstract meaning, such as that **m** refers to something called the mass of a body and **v** to its velocity at a specified instant of time, **t**. The equations can then be used to construct a vast array of abstract mechanical systems, for example, a two-body system subject only to mutual gravitational attraction. I call such an abstract system a *model*. By stipulation, the equations of motion describe the behavior of the model with perfect accuracy. We can say that the equations are exemplified by the model or, if we wish, that the equations are *true,* even *necessarily* true, for the model. For models, truth, even necessity, comes cheap.

The connection to the world is provided by a complex relationship between a model and an identifiable system in the real world. For example, the Earth and the Moon may be identified as empirical bodies corresponding to the abstract bodies in the model. The mass of the body labeled $m_1$ in the model may be identified with the mass of the Earth while the distance **r** in the model is identified with the distance between the center of the Earth and the center of the Moon. And so on. Then the behavior of the model provides a representation of the behavior of the real Earth-Moon system. For the purposes of understanding the relationship by which the model represents the real system, the concept of truth is of little value. A model, being an abstract object rather than something linguistic, cannot literally be true or false. We need another sort of relationship altogether.

Some friends of models invoke *isomorphism,* which is at least the right kind of relationship.[17] But isomorphism is too strong. The same considerations that show the strict falsity of presumed universal laws argue for the general failure of complete isomorphism between scientific models and real-world systems. Rather, models need only be similar to particular real-world systems in specified respects and to limited degrees of accuracy. The question for a model is how well it "fits" various real-world systems one is trying to represent. One can admit that no model fits the world perfectly in all respects while insisting that, for specified real-world systems, some models clearly fit better than others. The better fitting models may represent more aspects of the real world or fit some aspects more accurately, or both. In any case, "fit" is not simply a relationship between a model and the world. It requires a specification of which aspects of the world are important to represent and, for those aspects, how close a fit is desirable.

In this picture of science, the primary representational relationship is between individual models and particular real systems, e.g., between a Newtonian model of a two-body gravitational system and the Earth-Moon system. But similar models may be developed for the Earth-Sun system, the Jupiter-Io system, the Jupiter-Sun system, the Venus-Sun system, and so on. Here we have not a universal law, but the *restricted generalization* that various pairs of objects in the solar system may be represented by a Newtonian two-body gravitational model of a specified type. Restricted generalizations have not the form of a universal statement plus a proviso, but of a conjunction listing the systems, or kinds of systems, that may successfully be modeled using the theoretical resources in question, which, in our example, are Newton's equations of motion and the formula for gravitational attraction.

Other pairs of objects in the solar system cannot be well represented by the same sort of model, the Earth-Venus system, for example. Moreover, although one could in principle construct a single Newtonian model for all the planets together with the Sun, the resulting equations of motion are not solvable by any known analytical methods. One cannot even solve the equations of motion for a three-body gravitational system, one intended to represent the Earth-Jupiter-Sun system, for example. Here one must approximate, for example, by treating the influence of the Earth as an perturbation on the motion of the two-body Jupiter-Sun system. Such approximation techniques have been part of Newtonian practice since Newton himself, but have been largely ignored by the tradition that interprets Newton's equations of motion as expressing universal laws of nature.

It is typically said to be a major part of Newton's success that he "unified" the behavior of celestial and terrestrial bodies. The equations of motion used to build models of the Jupiter-Sun system may also be used to construct models to represent the behavior of balls rolling down an inclined plane, pendulums, and cannonballs. This was a considerable achievement indeed, but it hardly elevates the equations of motion to the status of universal laws. It had yet to be shown that similar models could capture the comings and goings of comets, and the fixed stars were beyond anyone's reach. What Newton had in 1687 were not God's all-encompassing laws for nature, but a broad, though still restricted, generalization about some kinds of systems that could be modeled using the resources he had developed. That he had fathomed God's plan for the universe was an interpretation imported from theology.

## 5. Principles versus Laws

It may reasonably be objected that focusing simply on Newton's equations of motion does not do justice to their role in the science of mechanics. They seem somehow to capture something fundamental about the structure of the world. One might express similar feelings about the Schrödinger equation in quantum mechanics. The problem is to capture this aspect of these fundamental equations without lapsing back into the language of universal laws.

An interpretive device that has considerable historical precedent would be to speak of Newton's *Principles of Motion* and the *Principle of Gravitational Attraction*. The title of his book, after all, translates as *The Mathematical Principles of Natural Philosophy*.[18] Whether or not thinkers in the seventeenth, or even eighteenth, century recognized any significant distinction between "laws" and "principles," we can make use of the linguistic variation. Principles, I suggest, should be understood as rules devised by humans to be used in building models to represent specific aspects of the natural world. Thus Newton's principles of mechanics are to be thought of as rules for the construction of models to represent mechanical systems, from comets to pendulums. They provide a *perspective* within which to understand mechanical motions. The rules instruct one to locate the relevant masses and forces, and then to equate the product of the mass and acceleration of each body with the force impressed upon it. With luck one can solve the resulting equations of motion for the positions of the bodies as a function of time elapsed from an arbitrarily designated initial time.

What one learns about the world is not general truths about the relationship between mass, force, and acceleration, but that the motions of a vast variety of real-world systems can be successfully represented by models constructed according to Newton's principles of motion. And here "successful representation" does not imply an exact fit, but at most a fit within the limits of what can be detected using existing experimental techniques. The fact that so many different kinds of physical systems can be so represented is enough to justify the high regard these principles have enjoyed for three hundred years. Interpreting them as universal laws laid down by God or Nature is not at all required.[19]

## 6. Necessity without Laws

Traditionally, it has been the supposed universality of laws of nature that has seemed to require their necessity. For, as Kant argued, how could a universal association be just a regularity? For an association to be truly universal, he thought, there must be something making it be so. Thus, denying the existence of genuine universal laws in nature makes it possible to deny the existence of necessity as well. But such denial is not required. It is also possible to deny the existence of universal laws of nature while affirming the existence of causal necessities.[20]

Consider a model of a harmonically driven pendulum of the sort that one would use to represent the motion of a pendulum on a typical pendulum clock. Solving the classical equations of motion for the period as a function of length (assuming that the angle of swing is small) yields the familiar result that the period is proportional to the square root of the length. Now this model provides us with a range of possible periods corresponding to various possible lengths. These possibilities are built into the model. But what of the real world?

Suppose the actual length of the pendulum on my grandfather clock is **L**. The model permits us to calculate the period, **T**. It also permits us to calculate a slightly greater period **T′** corresponding to a slightly greater length, **L′**. Suppose the clock is running slightly fast. I claim that turning the adjusting screw one turn counterclockwise would increase the length of the pendulum to **L′**, and this would increase the period to **T′**, so that the clock would run slightly slower. This seems to be a claim not about the model, but about the real life clock in my living room. Moreover, it seems that this claim could be true of the real life clock even if no one ever again touches the adjusting screw. These possibilities, it seems, are in the real physical system, and are not just features of our model.

There are, of course, many arguments against such a realistic interpretation of modal (causal) claims. Here I will consider only the empiricist argument that there can be no evidence for the modal claim that is not better evidence for the simple regularity relating length and period for pendulums. The inference to counterfactual possibilities, it is claimed, is an unwarranted metaphysical leap.[21] Moreover, I will not try to argue that this empiricist interpretation is mistaken; only that it is no less metaphysical than the opposing view.

I claim that by experimenting with various changes in length and observing changes in period one can effectively *sample* the possibilities that the model suggests may exist in the real system. That provides a basis for the conclusion that these possibilities are real and have roughly the structure found in the model. The empiricist argument is that the most one can observe is the actual relationship between length and period for an actual series of trials with slightly different initial conditions. So the issue is whether experimentation can reveal real possibilities in a system or merely produce actual regularities in a series of trials. Whichever interpretation one favors, one cannot claim that the latter interpretation is somehow less metaphysical than the former. It is just a different metaphysical view. I think the modal realist interpretation provides a far better understanding of the practice of science, but that is not something one can demonstrate in a few lines, or even a whole book.

# SIX

# The Cognitive Structure of Scientific Theories

## 1. Introduction

Within post–World War II analytic philosophy of science, assumptions about the structure of scientific theories have formed an essential part of the background for most other topics. Until around 1970, the "received view" (Suppe 1972) was that, for *philosophical* purposes, scientific theories are to be thought of as interpreted, formal, axiomatic systems. The axioms of a theory, on this account, are statements which, in principle, are either true or false. In addition, at least some of the axioms were typically taken to have the form of laws, understood as universal generalizations. On this account, then, scientific theories have the structure of an axiomatic, deductive system. This account of theories is now often referred to as the *statement view* (or *law-statement view*) of theories (Stegmüller 1976).

The qualifying phrase "for *philosophical* purposes" is important because no one ever argued that scientists, in their everyday work, actually use such a conception of theories. Rather, this conception of theories was to be employed in the philosophical reconstruction of particular scientific theories and in philosophical investigations into the general nature of science—its methods for empirically testing theories, its modes of explanation, the reduction of one theory to another, and so on. On this view there need be little connection between what philosophers of science say about the nature of theories and what historians, psychologists, or sociologists might learn about the use of theories in actual scientific practice.

The major alternative view of scientific theories has several names. It is sometimes called the "semantic view of theories," by way of contrast with the supposed "syntactic" character of theories on the received view. It is also called the "non-statement" view, the "predicate" view, or (as I now prefer) a "model-based" view of theories.[1] A simple way of describing a model-based view is to say that theories include two different sorts of linguistic entities. Some are predicates, which may have a quite elaborate internal structure, as, for example, the predicates "pendulum" or "two-body Newtonian gravitational system." Others are statements of the form "X is P" where X refers to a real world system and P is one of the predicates, as in the statement, "The Earth-Moon system is a two-body Newtonian gravitational system." The predicates, as such, have no truth values, but the associated statements do.

On a model-based view, there is no simple answer to the question "What is the structure of scientific theories?" The gross structure would seem to be that of a family of models (or predicates). But the models themselves may have a complex formal (mathematical) structure. And the linguistic structure corresponding to a predicate is not that of an axiomatic system, but of a definition. There is also a question as to whether the statements applying the models to the world should be included within the overall structure of the theory itself. If so, which ones? All of them? Some of them? Or none? If all, the theory changes with each new application. If some, why these rather than others? If none, then there are no empirical statements within the structure of a theory proper.

In spite of these questions, this way of characterizing the statement and model-based views of scientific theories makes the differences between them seem relatively formal, even trivial. Expressions that function as empirical laws on the statement view may reappear in the definitions of predicates on the model-based view. And there seem to be few if any significant empirical claims that could be formulated in one framework and not in the other. Advocates of the model-based view have, of course, offered a variety of reasons for thinking not only that the difference is not trivial, but that the model-based view is clearly superior. It is often argued, for example, that the model-based approach better supports formal inquiries into the "foundations" of particular theories. I will not rehearse these arguments here.[2]

My goal in this chapter is to explore a new reason for preferring a model-based approach. It is, in a nutshell, that adopting a model-based framework makes it possible to employ resources in cognitive psychology to understand the structure of scientific theories in ways that may illuminate the role of theories in the ongoing pursuit of scientific knowledge. This is part of a more general project of developing a comprehensive, naturalistic account of science as a human activity. I will begin by suggesting a general connection between a model-based understanding of scientific theories and research into the nature of concepts and categories by cognitive psychologists. Then I will review one influential line of psychological research which provides a basis for the subsequent exploration of the cognitive structure of a prototypical scientific theory, classical mechanics.[3] I will conclude by arguing that a concern with the cognitive structure of scientific theories is largely orthogonal to the sorts of foundational concerns with which the phrase "the structure of scientific theories" has typically been associated.

## 2. Concepts and Models

Surveying work in the cognitive sciences, particularly cognitive psychology, one cannot miss a large body of work on the nature of concepts, categories, and classification going back at least to Jerome Bruner's work in the 1950s (Bruner, Goodnow, and Austin 1956). Particularly prominent is a series of investigations into "natural categories" carried out by Eleanor Rosch and various collaborators in the decade between the mid-1970s and mid-1980s.

Philosophers of science are accustomed to thinking of categorization as preliminary to theoretical science, as natural history is preliminary to the theory of natural selection. At best, it is commonly thought, classification leads to empirical generalizations which are then systematized in theories. So, there is thought to be some relationship between categorizing and theorizing, but not one that promises much insight into the nature of theorizing. This reaction, however, is itself grounded in the view that theories are sets of statements, perhaps sets of universal generalizations. Viewed through the lens of a model-based understanding of theories, the connection between theories and categories is potentially much more fundamental. Even couched in linguistic terms, a model functions as a predicate, as a model of a pendulum gives content to the predicate "pendulum" in the open sentence "X is a pendulum." So there is initially at least the possibility that some of what anthropologists, psychologists, and linguists have discovered about the structure of naturally occurring concepts might be carried over to the study of the families of models which, on the model-based approach, are a major component of most scientific theories. I will now argue that this possibility is not only realized, but fruitful as well.

Before proceeding, however, I must note one point on which there is near unanimity among cognitive psychologists. This is the inadequacy of the "classical model" of categories, namely, the view that membership in a category is determined by a set of necessary and sufficient features (Smith and Medin 1981). Since much work in the philosophy of science presupposes a set-theoretical reconstruction of scientific theories—and set membership is, of course, determined by necessary and sufficient conditions—there is here a major divergence between the approach of philosophers of science and that of cognitive psychologists.[4] I will attempt a reconciliation of these two approaches at the end of this chapter. The reasons cognitive psychologists give for rejecting the classical model of concepts include those advanced by Wittgenstein fifty years ago, namely, that because there are so many unclear cases, one simply cannot find necessary and sufficient conditions for most concepts. The main reason, however, is more positive. It is that typical concepts exhibit structural features that simply are not accounted for by the classical model.

## 3. Concepts and Categories: *Basic Color Terms*

A good place to begin is with the investigations of Berlin and Kay (1969) on basic color terms. These investigators explicitly set out to refute what they called "the doctrine of extreme linguistic relativity" (1969, 1), a doctrine associated with the work of Sapir and Whorf (Whorf 1956). Applied to color terms, the relativistic doctrine had been that the linguistic division of the visible spectrum in different languages is totally arbitrary, the product solely of cultural circumstances and history. Berlin and Kay, to the contrary, argued that, among all existing languages, there are only *eleven* basic color terms. Languages differ in the number of basic color terms they exhibit, from two to all eleven, but even which combinations of colors are represented in various languages is not arbitrary.

Very briefly, the research leading to these conclusions consisted of two stages. They first devised a set of eight criteria for determining the "basic color terms" of any language. For example, a basic color term must be "monolexemic," that is, not composed of meaningful parts from which the meaning of the whole term might be determined. By this criterion, for example, "light blue" cannot be a basic color term, but "blue" could be.

They then presented native speakers of twenty different languages with a standard color chart "composed of 320 color chips of forty equally spaced hues and eight degrees of brightness, all at maximum saturation, and nine chips of neutral hue (white, black and grays)" (p. 5). For each basic color term, B, in the speaker's language, native speakers were asked to indicate on the chart (1) "all those chips they would under any circumstances call B" and (2) "the best, most typical examples of B" (p. 7).

For speakers of the same language, even the same speaker at different times, the *boundaries* of application for color terms were quite variable. But the *foci* were remarkably fixed, typically varying by no more than one chip in any direction. It is the existence of clearly defined foci that permits the identification of basic color terms with single "focal colors." These findings were confirmed by secondary sources regarding an additional seventy-eight languages.

*Natural Categories*

Having herself worked on color categories, Rosch (1973a) wondered whether other natural categories corresponding to concepts like "circle" or "bird" or "furniture" might not exhibit a similar graded structure. They do. Much of the evidence consists of experiments performed using the cognitive psychologist's favorite experimental animal, the university undergraduate. These sorts of experiments have been replicated in many contexts with similar results.

One kind of evidence comes from free response experiments in which subjects are asked simply to list examples of a given category. Thus, for the category "bird," a particular subject's list might include "robin" and "sparrow." Figure 6.1 shows some response frequencies for various types of birds. The responses have been selected so as to exhibit a roughly uniform spread in response frequency from most frequent to least frequent. Alongside are direct ratings of "exemplariness" on a seven point scale. Here subjects were asked explicitly to rate how well the particular example fits their own idea of what the general concept represents. How well, for example, does a sparrow fit their own idea of what a bird should be? There is a clear correlation between the two scales (Rosch 1973b).

Using the free response data, one can make a rough division between "central" and "peripheral" instances of a concept. Rosch and her collaborators then performed classic reaction time experiments in which the subject was presented, on a computer screen, with sentences like "A robin is a bird." or "A duck is a bird." or "A robin is a fruit." The subject was instructed to respond "true" or "false" as rapidly as possible by pressing appropriate keys. The prediction was that the reaction times for true sentences would be shorter for cen-

6.1 Free response frequencies compared with direct ratings of "exemplariness" for members of the category "bird"

| Category | Member | Response Frequency | "Exemplariness" rank |
| --- | --- | --- | --- |
| Bird | Robin | 377 | 1.1 |
| | Eagle | 161 | 1.2 |
| | Wren | 83 | 1.4 |
| | Chicken | 40 | 3.8 |
| | Ostrich | 17 | 3.3 |
| | Bat | 3 | 5.8 |

Source: Rosch (1973b).

6.2 Reaction times for judging the truth or falsity of subject-predicate statements employing either central or peripheral instances of a category

| | Response | | | |
|---|---|---|---|---|
| | True sentences | | False sentences | |
| Category member | Reaction time (msec) | Error Proportion | Reaction time (msec) | Error Proportion |
| | *Adults* | | | |
| Central | 1011.67 | .028 | 1089.94 | .024 |
| Peripheral | 1071.45 | .071 | 1115.52 | .032 |
| | *Children* | | | |
| Central | 2426.45 | .056 | 2692.40 | .038 |
| Peripheral | 2703.45 | .228 | 2799.30 | .029 |

Source: Rosch (1973b).

tral cases than for peripheral cases. In fact, as shown in figure 6.2, there is roughly a six percent difference, which is statistically significant, in the predicted direction. Also, the proportion of errors is over twice as great for true sentences with peripheral as opposed to central instances (Rosch 1973b). The data are even stronger, although more difficult to interpret, when the subjects are children rather than adults. In any case, the presumption is that these differences in reaction times and error rates reflect a real difference in how the concepts are cognitively structured.[5]

Graded membership in ordinary categories is often referred to as demonstrating a "horizontal," or "within category," structure to concepts. Rosch also claimed that there is a "vertical," or "across category," structure as well. Consider the hierarchy: Living Thing, Mammal, Dog, Collie, Lassie. Logically the relationship between levels is simply one of inclusion. Rosch argued that the intermediate level, "Dog," is *basic*. What makes this level "basic," she claimed, is that this is the most general level for which there is a high degree of similarity among members. Members of superordinate levels typically exhibit a much lower degree of similarity, while the gain in similarity when moving to a subordinate level is generally small. (Rosch 1978).[6]

Similarity, of course, has many measures. One is simply number of shared salient features. Another is strength of correlation among features. Another is visual similarity. Rosch claimed that objects in a basic-level category exhibit greater similarity than those in a superordinate category in all these ways, and in many other ways as well. For example: (1) Subjects were much more adept at identifying superimposed standardized silhouettes of two basic-level objects than two superimposed silhouettes of different basic-level objects belonging to the same superordinate category. The superposition of an easy chair and a desk chair, for example, is much more recognizable as a chair than a superposition of a couch and a bed is as either (Rosch et al. 1976). (2) Other studies show that, at any age, basic-level terms are more quickly learned than either superordinate or subordinate terms. And reaction time studies show that objects are more quickly identified visually using basic-level terms than superordinate or subordinate terms. Thus an object is more quickly identified as a chair than either as a piece of furniture or as an easy chair (Rosch et al. 1976).

Inspired by Rosch's work, cognitive psychologists developed various theories which might account for her results, particularly the existence of graded rather than classical structures for individual categories. The main theories of graded structure have been *similarity theories*, of which there are several varieties. In a typical account, individual real objects are classified as belonging to a particular category by their degree of similarity with an idealized object, sometimes called a "prototype," which is defined classically by a set of necessary and sufficient features. The degree of similarity might then be determined by the number of shared features, with more salient features being given greater weight. The features for "bird," for example, might include "has wings," "flies," and "eats seeds," with having wings being the most salient. In a variant account, membership in the category might be determined by degree of similarity with one or more specific types of object, sometimes called "exemplars," within the more general category. Thus, for example, if "robin" is an exemplar for "bird," categorization as a bird could be determined by degree of similarity with an exemplary robin.[7]

There seems now to be considerable dissatisfaction with all similarity based accounts of concepts. One line of criticism is that similarity accounts disregard information that seems to be part of the concept, such as that contained in correlations among relevant properties. For example, it is argued that anyone who knows what birds are knows that small birds are more likely to be songbirds than are large birds. It is very difficult to build this information into a similarity theory. A deeper objection is that similarity theories leave unexplained why some properties are relevant for a category and others are not. In versions which allow some properties to count more than others, why should the weights be what they are? What, it is asked, holds the concept together?[8]

As a remedy for these difficulties, several cognitive psychologists have recently proposed a "knowledge-based," or "theory-based" account of concepts. The idea here is that there are underlying principles which create a network of causal and explanatory links which both hold individual concepts together and provide connections with related concepts.[9] This latest work, however, suffers from a tacit acceptance of a law-statement picture of theories. As will soon become clear, the account I am developing here amounts to a theory-based account of categorization which identifies the core of a scientific theory with an array of models.[10]

*Idealized Cognitive Models*

In addition to Rosch's work, my account draws inspiration from the theory of "Idealized Cognitive Models" (ICMs) developed by the self-described "cognitive linguist" George Lakoff (1987). According to Lakoff, all accounts of categorization which compare features to measure similarity are fundamentally misconceived. Rather, he claims, people operate with sets of ICMs, and it is relationships among whole ICMs, not similarity based on individual features, that produce the gradations in classifying behavior that Rosch uncovered.

One of Lakoff's favorite examples is the concept "mother" in contemporary American culture. This concept exhibits what Lakoff calls a *radial structure,* that is, a structure which "radiates" out from a central, or "focal," model. But the less central models are not just logical subcases of the central model. Nor are the less central models produced simply by forming Boolean combinations of features of the central model. They are extensions which may be motivated by all sorts of contingent circumstances.

Pursuing this example a bit further, Lakoff characterizes the focal concept "mother" as a person who (1) is and always was female, (2) gave birth to the child, (3) supplied half the child's genes, (4) nurtures the child, (5) is married to the child's father, (6) is one generation older than the child, (7) is the child's legal guardian. A "stepmother," by contrast, is characterized as a person who (1) did not give birth to the child, (2) did not supply any of the child's genes, (3) is currently married to the child's father; while a "foster mother" is characterized as a person who (1) did not give birth to the child, (2) is paid by the state to nurture the child (1987, 83). In this example, Lakoff claims, the graded structure results from overlaps and other connections among the various models, with the focal model being at the center of a radiating structure. It is the focal model that provides whatever unity the whole family of models may exhibit.

We need not pursue the details of this example any further here. All I want to take away is the general suggestion that underlying Rosch's graded structures are families of models exhibiting various relationships yielding judgments about which models are more central and which more peripheral. I wish to move on to see whether Rosch's results, and Lakoff's suggestions, can shed any light on the families of models that, according to a model-based approach, constitute the core of *scientific* theories.

## 4. The Graded Structure of Concepts in Classical Mechanics

I know of no empirical studies of graded structure carried out using as categories simple mechanical systems. But such studies have included geometrical shapes (circles, squares, triangles) (Rosch 1973a) as well as vehicles (cars, trains, airplanes) and weapons (guns, bombs, and clubs) (Rosch and Mervis 1975). It is therefore hard to imagine that the usual sorts of experiments would not yield more and less central cases of such ordinary mechanical systems as pendulums. Figure 6.3, for example, exhibits an intuitively plausible sequence of diagrams representing real pendulums ranging from central (a, b) to

# THE COGNITIVE STRUCTURE OF THEORIES | 107

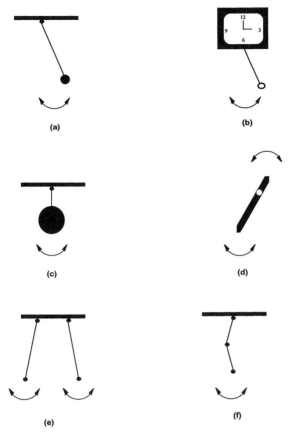

6.3  Diagrams representing real pendulums ranging from central (a, b) to peripheral (e, f) cases.

peripheral (e, f) cases. One might argue with particular placements in the sequence, but not, I think, with the overall progression from central to peripheral cases.

It is a reasonable expectation that judgments regarding central versus peripheral examples of pendulums might differ depending on one's knowledge of classical mechanics. "Novices," that is, people with little or no formal training in classical physics, might well produce a different ordering than "experts," such as physics professors. Unfortunately, studies of the "novice-expert shift" in physics (Chi, Feltovich, and Glaser 1981; Larkin, McDermott, Simon, and Simon 1980) have not focused on simple tasks of classification that would

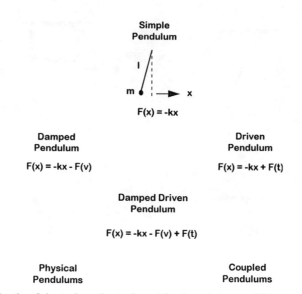

6.4   A family of classical mechanical models of pendulums exhibiting a radial structure.

reveal graded structure. Still, there seems little reason to doubt that the judgments of experts would exhibit some progression from central to peripheral cases for mechanical systems like pendulums. Assuming that is so, can we find a rationale for a graded structure in the families of *models* of classical mechanics that could be used to represent such cases? In particular, can we find something like a focal model for pendulums corresponding to Lakoff's central ICM for mother? I think we can.

At this point I will assume that one can identify families of classical mechanical models. Substantiating this assumption is not necessary to my argument here, which is conditional. The argument will be that adopting a model-based approach makes it possible to apply Rosch's analysis in a way that usefully increases our understanding of how theories are structured in scientific practice. And that is a reason for preferring a model-based account.

Figure 6.4 shows a family of models of pendulums. The focal model is clearly the *simple pendulum,* for which the horizontal restoring force (assuming small angular displacements) is a linear function of the horizontal displacement, **x**. Models of either damped or driven pendulums require additional forces, negative for damped and positive for driven pendulums. The damped and driven pendulum has both types of additional forces. Coupled pendulums are a type of driven pendulum, where the driving force is another pendulum which is part of the whole system. There is no external source of energy, but there could be. That would be yet another type of model. Physical pendulums can be analyzed as regular pendulums if one first calculates the moment of inertia around the pivot point and then determines what would be the length of an equivalent simple pendulum, which, finally, yields the frequency of oscillation of the physical pendulum. And of course physical pendulums can also be damped or driven, or both.

What is the basis for the ordering shown in figure 6.4? It is the obviously increasing complexity of those models below the simple pendulum. For example, even the simplest model of a damped pendulum would add to the linear restoring force an additional restoring force which is a linear function of velocity. But other, more complex, functions of velocity are also possible. One does not need an abstract definition of complexity (or simplicity) to judge that models incorporating forces which are an arbitrary function of velocity or time are more complex than ones in which the only force is a linear function of position.

It is worth insisting that the complexity in the models appealed to here is not a mere "descriptive" complexity. That is, it is not a complexity in the *equations* that matters; it is a complexity in the *models* themselves, a complexity introduced into the models in order better to represent actual physical systems. This complexity is reflected in the equations as I have presented them, but that is not its source. Other notations could fail to reflect the underlying complexity of the forces involved.

Finally, the fact that a damping force can be any function of velocity shows that the graded structure of the set of models for pendulums goes much deeper. The central model for damped pendulums employs a linear function of velocity with a single coefficient of friction. But other models, incorporating other functions of velocity, radiate out from this model, creating a further graded structure within the category of damped pendulums. This is of course true for driven pendulums as well.

## 5. The Basic Level of Models in Classical Mechanics

Figure 6.5 exhibits a multiple hierarchy of families of models of classical mechanics. One may think of it as a partial "model map" of classical mechanics.[11] The pictures at Level V are to be interpreted as "visual models" corresponding to particular subfamilies of models indicated at Level IV.

Looking at Levels III and IV, we note that pendulums are not the only species of harmonic motion. If anything, a bouncing spring is an even more central case of harmonic motion than a pendulum. Figure 6.6 provides a sequence of real-world examples moving from intuitively central (a, b) to intuitively peripheral (e, f) cases. All are instances of simple harmonic oscillators, so the corresponding models are not distinguished by the existence of additional forces in some models. The rotational examples, however, introduce the

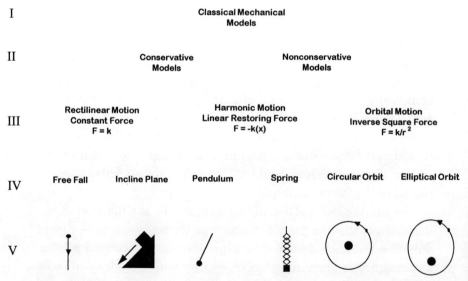

6.5  A partial "model map" of classical mechanics exhibiting a multiple hierarchy of models.

# THE COGNITIVE STRUCTURE OF THEORIES | 111

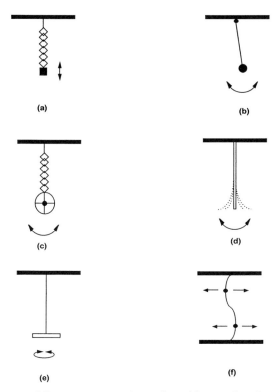

6.6 A sequence of diagrams representing real-world examples of simple harmonic oscillators ranging from central (a, b) to peripheral (e, f) cases.

complexity that it is not just mass that matters, but the moment of inertia, which is a function of the distribution of the mass, e.g., in a disk as opposed to a ring. Similar complexities inhabit the models for a vibrating cantilevered beam or a section of a vibrating string. So there are features of the models themselves that support intuitive judgments concerning the graded structure of harmonic systems as a category. But explaining the existence of a graded structure at this level is not my main concern here. Mainly I want to argue that Level IV, the level of pendulums and springs, is a Roschian *basic level* in the categorical structure of classical mechanics.

First off, at this level the members of the various categories satisfy the Roschian criterion of *visual similarity;* that is, the members of the category typically appear visually more similar with each other

6.7 Twenty elementary problems in classical mechanics classified according to the general principles required to solve them

| Surface Structure | Forces | Energy | Momentum |
|---|---|---|---|
| Pulley with hanging blocks | | 20 | |
| | 11 | 19 | |
| | 14 | 3 | |
| Spring | | 7 | |
| | 18 | 16 | 1 |
| | | | 17 |
| | | 9 | 6 |
| | | 3 | 3 |
| Inclined plane | 14 | 5 | 5 |
| Rotational | 15 | | 2 |
| | | | 13 |
| Single hanging block | 12 | | |
| Block on block | 8 | | |
| Collisions (Bullet-"block" or Block-block) | | | 4 |
| | | | 6 |
| | | | 10 |

Source: Chi, Feltovich, and Glaser (1981).

than with members of other categories at the same level. Pendulums swing, springs bounce, and torsion bars twist. Moreover, as is the case when moving up from "chair" to "furniture," much of this visual similarity is lost when one moves up from pendulums and springs to the more abstract, superordinate level of simple harmonic oscillators.

Another Roschian indicator of the basic level is that categories at this level are the first to be learned. Here there is some relevant empirical evidence. Some years ago, M. T. H. Chi and several collaborators studied differences in the way textbook problems in classical mechanics are categorized and solved by novices as opposed to experts (Chi, Feltovich, and Glaser 1981). They selected twenty problems from a standard text, representing roughly a half dozen different categories at my Level IV, as shown in figure 6.7. These problems were selected so as to represent roughly equal numbers of problems in which the standard method of solution involved (i) Newton's Second Law, (ii) conservation of energy, and (iii) conservation of momentum. Some problems required two of these principles and

6.8 The classification of the twenty problems produced by a "novice"

| | | |
|---|---|---|
| Group 1: | 2, 15 | "Rotation" |
| Group 2: | 11, 12, 16, 19 | "Always a block of some mass hanging down" |
| Group 3: | 4, 10 | "Velocity problems" (collisions) |
| Group 4: | 13, 17 | "Conservation of Energy" |
| Group 5: | 6, 7, 9, 18 | "Spring" |
| Group 6: | 3, 5, 14 | "Inclined plane" |

Source: Chi, Feltovich, and Glaser (1981).

6.9 The classification of the twenty problems produced by an "expert"

| | | |
|---|---|---|
| Group 1: | 2, 13 | "Conservation of Angular Momentum" |
| Group 2: | 18 | "Newton's Third Law" |
| Group 3: | 1, 4 | "Conservation of Linear Momentum" |
| Group 4: | 19, 5, 20, 16, 7 | "Conservation of Energy" |
| Group 5: | 12, 15, 9, 11, 8, 3, 14 | "Application of equations of motion" (F = MA) |
| Group 6: | 6, 10, 17 | "Two-step problems: Conservation of Linear Momentum plus an energy calculation" |

Source: Chi, Feltovich, and Glaser (1981).

some could be solved in more than one way. Subjects were asked to group the problems into categories and also state the basis for their categorization.

Figure 6.8 shows the result for a "novice," that is, a student who had completed one course in mechanics. The constructed categories are mostly at my Level IV.[12] By contrast, an "expert," a physics professor, classified the problems according to the general principles that would be employed in their solution, as shown in figure 6.9. The task, after all, was to classify *problems,* not models.

Chi and her collaborators described these results as supporting the view that novices classify problems on the basis of surface features while experts classify on the basis of underlying physical principles. My explanation is different, though not incompatible with their description. I think the novices were operating at the basic level in the hierarchy of categories, as would be expected on Rosch's analysis. The experts were using superordinate categories at my Levels II and III. So, on my account, part of becoming an expert in a science like mechanics is learning to categorize systems at levels of abstraction *above* the Roschian basic level. This is in general agreement with other research on the novice-expert shift (Larkin 1985).[13]

## 6. Higher Level Categories

Turning to Level III itself, the three examples on my model map are all pretty central examples of Newtonian models. But it is easy to imagine less central models. Consider, for instance, a polynomial force law, or an inverse seventh power force law. I think these would be regarded as peripheral models by anyone familiar with classical mechanics. And one can find a rationale for this judgment in the fact that these latter models are complicated in ways the central models are not. I suspect that here there is an additional reason for regarding such models as peripheral. They have few applications. That is, few systems one meets in the real world call for such models.

It is important to realize the respects in which the set of possible categories at Level III is and is not well defined. These categories are determined by the nature of the force function. But Newton's laws of motion impose few limits on what might count as a force function. So the class of force functions is potentially nearly as extensive as the class of all possible mathematical functions from the real numbers onto the real numbers. In the context of classical logic and mathematics, this class as a whole is regarded as mathematically well defined. However, there can be no algorithm for generating all members of this class. That can be done only for more restricted sets of functions. So there can be no question of explicitly surveying all the possible categories of Newtonian models at Level III.

The situation is similar at Level II. A *conservative* system can be characterized as one for which the total energy (or, alternatively, the Hamiltonian) is constant in time, no matter what the energy function happens to be. All other systems are *non-conservative*. So the distinction between conservative and non-conservative systems is in general well defined. But there is no way to construct the set of all possible conservative systems. Finally, at Level I, a Newtonian system is defined as one for which Newton's three laws of motion are satisfied. That definition is independent of the particular form of the force function. But there is no way further to define a more specific model of a Newtonian system without specifying a force function.

In the axiomatic approach to understanding classical mechanics, Newton's laws of motion play the role of axioms, typically in the form of universal generalizations. On my account, Newton's Laws, particularly the second, provide the formal structure both for superordinate categories of models at Levels II and III and for particular models at Level V. They do not take the form of universal generalizations. They function more like recipes for constructing models than like general statements. The relevant empirical generalization is that the behavior of many types of real world systems can in fact successfully be represented by models constructed using this recipe.

## 7. Formal Structure and Cognitive Structure

I would like now to return to the potential conflict between the rejection by cognitive psychologists of the "classical" account of concepts and its general acceptance by philosophers of science. From a philosophical perspective, one might express the conflict as follows: Rosch's analysis concerns natural categories used in everyday life. Even her investigations into concepts like "circle" and "square" concern the everyday uses of these concepts, not their technical, mathematical uses. But mature scientific theories, like those of physics, are formulated in the language of mathematics. And mathematics reduces (more or less) to set theory, which presupposes the classical model of concepts. So, it is concluded, the classical model of concepts must be right for scientific theories. Whatever the right account of everyday concepts might be, it is simply irrelevant for an understanding of scientific concepts.

In favor of this objection is the fact that every Newtonian model is well-defined along classical lines. A necessary and sufficient condition for a Newtonian model to be a simple harmonic model, for example, is that a mass be subjected to a linear restoring force. The behavior of any such model can be determined by straightforward mathematical analysis. So any particular Newtonian model has a fully specified formal structure. This is the structure that engages the creative efforts of theoretical scientists and, presumably, philosophers pursuing the "foundations" of special sciences. This structure, however, is completely *internal* to the models under investigation.

The Roschian horizontal and vertical structures, on the other hand, are generated by relationships between and among models, and by relationships between models and human agents. These structures are *external* to the individual models, even though they may to some extent be explained by differences in the internal structure of the particular models involved. Being a peripheral model, for example, makes sense only by contrast with central models. In brief, the "classical" structures are found *within* models; the levels and graded structures within levels are a function of relationships *among* models.

Axiomatic presentations of classical mechanics, the canonical mode of presentation on the received view of scientific theories, miss the structure exhibited in my model map (fig. 6.5). A typical axiomatic presentation, for example, begins at Level I with Newton's laws and then adds the gravitational law, which incorporates the conservative (Level II) inverse square force law (Level III). Only by adding specific initial conditions does one reach Levels IV or V. So the axiomatic presentation implicitly selects a particular path down through the hierarchy of models. No one would deny that picking a different force law would result in different theorems being derived. But, within the received framework, one would not see this as selecting a different path though a hierarchy of models.

The Roschian vertical structure, a basic level of models with superordinate and subordinate levels of models, presents a very different picture. What makes the basic level basic is not the internal structure of the models themselves, but the nature of various cognitive interactions between human agents and the real systems these models represent. These systems appear visually similar to human observers, and are the systems to which newly acquired theoretical models are first applied, and by reference to which they are first understood. I would guess that most of the real systems involved in Kuhn's exemplary problem solutions (Kuhn 1962, 1974) are systems corresponding to basic-level models. Looking more generally at the history of science, the anecdotal evidence, at least, is encouraging. We all know that Galileo voted a great deal of effort studying falling bodies, motion on inclined planes, and pendulums. Newton studied the same sorts of cases, particularly projectiles and the orbital motion of the Moon. All these studies were at the Roschian basic level, as one would predict if one assumes that the cognitive entry level for individual learning roughly correlates with the entry level for historical discovery.[14]

Indeed, although it is impossible to pursue the subject here, I would suggest that to understand the meaning or reference of theoretical terms, one should look first to how novice scientists learn to deploy basic-level models. One would find, I think, that there is no special problem about the meaning of theoretical terms. That was a pseudoproblem generated by the assumption that the essence of a scientific theory is to be found in the formal structure of its highest-level "laws." Having made that assumption, one then faced the "problem" of understanding how the formal structure could carry empirical significance. It was to solve this apparent problem that such artificial devices as meaning postulates and correspondence rules were invented. If one begins, however, with the use of basic-level models, and then works up to superordinate levels, there is no problem to be solved. One does not have to worry about how to put empirical significance into a formal structure if one avoids the initial leap of abstraction away from the meaning that was there all along at the basic level.

## 8. Conclusion

Advocates of a model-based understanding of scientific theories have paid insufficient attention to the overall structure of the families of models that, on that account, are the primary constituents of theories. Whenever the issue has arisen, it has mostly been addressed in terms of the internal, formal structure of the constituent models.[15] Relating the models in philosophers' model-based accounts of theories with the concepts in cognitive scientists' accounts of categorization suggests a structure to families of models far richer than has commonly been assumed. Families of models may be "mapped" as an array with "horizontal" graded structures, multiply hierarchical "vertical" structures, and local "radial" structures. These structures offer the promise of important implications for understanding how scientific theories are learned and used in actual scientific practice.

# SEVEN

# Visual Models and Scientific Judgment

## 1. Introduction

When reading scientific papers or watching presentations by scientists, nothing is more obvious than the use of *visual* modes of presentation for both theory and data. This not a new phenomenon, although it has been emphasized recently by the development of computer graphics. One finds a widespread use of various visual devices going back to before the scientific revolution (Baigrie 1996). And Newton's *Principia,* for example, is full of diagrams used in his geometrical demonstrations. But why should anyone be particularly interested in the use of pictures and diagrams in science? Specifically, why should a *philosopher of science* be interested in this particular aspect of the practice of science?

It is my view that studying visual modes of representation in science provides an entrée into fundamental debates within the philosophy of science, as well in related fields such as the history, psychology, and sociology of science. I will begin by indicating the nature of these debates and pointing out the relevance to these broader issues of the role played in science by visual modes of presentation. In the latter part of the chapter, I will use some diagrams that played a central role in the twentieth-century revolution in geology in order to illuminate these general themes.[1]

## 2. General Issues

Within the English-speaking world, the Logical Empiricist image of science, and the projects it generated, dominated philosophical thought about science for a generation following World War II. Two fundamental aspects of this image are relevant here. First, scientific knowledge consists primarily of what is encapsulated in scientific theories, and theories are ideally to be thought of as interpreted axiomatic systems. It follows that the primary mode of representation in science is *linguistic* representation. Second, the reasoning which legitimates the claims of a particular theory as genuine knowledge has the general character of a logic. That is, there are rules which operate on linguistic entities, yielding a "conclusion" or some other linguistic entity, such as a probability assignment.

In the framework of Logical Empiricism, then, there can be no fundamental role in science for nonlinguistic entities like pictures or diagrams. Such things might, of course, play some part in how scientists actually learn or think about particular theories, but unless their content is reduced to linguistic form, they cannot appear in a *philosophical* analysis of the content or legitimacy of any scientific claims to knowledge.

Like so many other aspects of post–World War II Western culture, the Logical Empiricists' picture of science began to blur in the decade of the 1960s. A major stimulus for change, and focus for opposing views, was Thomas Kuhn's *Structure of Scientific Revolutions* (1962). The initial rejection of Kuhn's views by philosophers of science was to be expected because he rejected the major assumptions of Logical Empiricism. According to Kuhn, for example, general statements organized into axiomatic systems play little role in the actual practice of science. There is thus little to be learned about science by reconstructing theories in a Logical Empiricist mold. Moreover, the relative evaluation of rival paradigms is not something that can be reduced to any sort of logic. It is fundamentally a matter of choice by scientists acting as individuals within a scientific community. For Kuhn, science is primarily a puzzle-solving activity. Scientific revolutions are the result of many individual scientists making the judgment that a particular type of puzzle, or way of approaching puzzles, is no longer fruitful, and that another approach provides a more promising basis for further puzzle-solving activities.

Kuhn himself did not highlight the role of visual or other nonpropositional modes of representation in science. Indeed, he avoided talk about representation. I surmise that was largely because he, like most everyone else, thought of representation in propositional terms, and that leads immediately to the concept of truth. His picture of science as a puzzle-solving activity was meant to be an alternative to the view of science as producing truths. Moreover, his emphasis on the incommensurability of terms in the languages of rival paradigms shows his tendency to think of scientific knowledge in linguistic categories.[2]

Nevertheless, Kuhn's approach to understanding science at least opened the door to consideration of nonlinguistic representational devices in the practice of science. This was not just because his account was historical, but because it was *naturalistic*. He was trying to explain how science works in terms of naturalistic categories like the psychological makeup of individual scientists and the social interactions among scientists in communities. Thus, whether nonpropositional devices like diagrams and graphs play a significant role in science is something to be determined empirically by examining actual cases of science in action.

Philosophers were initially quick to charge Kuhn with having fallen into epistemological relativism, a charge he personally struggled to avoid.[3] But beginning in the mid-1970s, several groups of European *sociologists* of science have pushed the relativistic aspects of Kuhn's views to their logical conclusion. The slogan of these schools is that science is a *social construct*. The import of the slogan is most quickly grasped by reflecting on the extent to which *society* is a social construct. There is, for example, nothing in the nonhuman universe that requires representative democracy, an independent judiciary, separation of church and state, or any other of the fundamental structures of American society. These are historically conditioned social constructs. Science, it is claimed, is no different. It follows that the world view of those we call "primitive" is in no objective way inferior to ours. It is just different. The only thing special about our scientific world view is that it is ours.[4]

Significantly, relativist sociologists of science were among the first to investigate the role of pictures, diagrams, and other nonpropositional forms of representation in science. Their aim has been to show how images are created and deployed in the social construction of scientific knowledge. The initially plausible view that these various images somehow picture reality is thereby "deconstructed."

There are more radical and less radical strains within the constructivist camp. A less radical view is to admit that scientists intend their theories to represent the world and often believe that they have succeeded. It is just that close sociological and anthropological analysis reveals that the intentions are not fulfilled and the beliefs mistaken. A more radical view is that science is not really a representational activity after all. In the twentieth century, painting clearly moved from being essentially representational to allowing forms that are not representational at all. We now have pictures that are not pictures of anything, and were never intended to be. So, it might be claimed, science is now (and maybe always has been) nonrepresentational. Our theories don't picture anything.[5]

My view is that what is needed is a middle way between philosophical positivism and sociological relativism, both of which, in very different ways, deny any genuine representational role for visual images in science. Examining visual modes of theorizing and evaluating data is part of a strategy for developing the desired middle way. Since images could not literally be true or false, this strategy avoids raising questions about the nature of truth. It thus makes possible the pursuit of a naturalistic theory of science which goes beyond puzzle solving to explore ways in which visual models might genuinely represent the real world and be correctly judged to do so.

As just indicated, there are several major parts to the overall program of developing a naturalistic middle way. A major task, for example, is simply to understand the various ways images and other nonpropositional devices can be used to represent the world. Here I will approach this task only to the extent of pointing out how a model-based understanding of scientific theories makes it possible to treat things like diagrams and scale models on a par with the more abstract theoretical models that, on this account, form the core of any scientific theory. The main focus of this chapter will be on explaining how pictorial presentations of data can be used in judging the relative representational adequacy of visually presented models of the world. Or, to put it in more traditional terms, I want to present a theory of scientific reasoning in which visual presentations of both data and theory can play a significant role. The 1960s revolution in geology provides a particularly rich context for just such a presentation.

## 3. Models and Theories

For a generation now, a number of philosophers of science have been developing an alternative to the Logical Empiricist account of scientific theories. This account has several names. It is sometimes called the "semantic view of theories," by way of contrast with the supposed "syntactic" character of theories on the received view. It is also called the "non-statement" view, the "predicate" view, or (as I now prefer) the "model-based" view of theories.[6]

On my understanding of a model-based approach to scientific theories, the predicate "pendulum," as it appears in classical mechanics, does not apply directly to real-world objects like the swinging weight in the grandfather clock that stands in my living room. It applies, rather, to a family of idealized models, the central example of which is the so-called "simple pendulum." A simple pendulum is a mass swinging from a massless string attached to a frictionless pivot, subject to a uniform gravitational force, and in an environment with no resistance. This is clearly an ideal object. No real pendulum exactly satisfies these any of these conditions. So no real pendulum is a simple pendulum as characterized in classical mechanics. And the same is true for more complex types of classical pendulums: damped pendulums, driven pendulums, and so on.

So what is the relationship between the idealized model pendulums of classical mechanics and real swinging weights? It is, I suggest, like the relationship between a *prototype* and things judged sufficiently similar to the prototype be classified as of that type. And how are such judgments made? After all, any two objects (idealized or not) are similar to each other in indefinitely many ways. Which features count for judgments of similarity to the prototype, and why do some

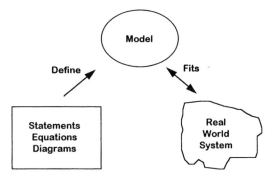

7.1  Relationships among language, models, and objects in the real world.

features count more than others? Here there are no simple answers. To some extent the models themselves provide guidelines for the relevant similarity judgments. The main dynamical variable in any model of a pendulum is the period of oscillation. It is a characteristic of the models that the mass of the bob is irrelevant to the period. Only the length of the suspension, plus the gravitational force, matters. So the mass of the bob should be relatively unimportant in classifying some real swinging weight as a simple pendulum. So should its shape.[7]

Figure 7.1 is an attempt to picture the relationships among language, models, and objects in the real world. The main point for present concerns is that, on this view of scientific theories, the primary representational relationship is not the truth of a statement relative to the facts, or even the applicability of a predicate to an object, but the similarity of a prototype to putative instances. This is not a relationship between a linguistic and a nonlinguistic entity, but between two nonlinguistic entities. Once this step has been taken, the way is clear to invoke other, less abstract, nonlinguistic entities to play a similar role.

4. Crucial Decisions

The idea of a "crucial experiment," as expounded, for example, by Francis Bacon, was a major cultural achievement of the Scientific Revolution. It deserves to be ranked along with such other achievements as the calculus, the telescope, and the air pump. How it came to have that status is a difficult historical question. Of course, with the hindsight of three centuries, we know that the role of crucial experiments has often been exaggerated, and that the designation of an experiment as "crucial" often comes long after the fact. But this only shows that the idea of a crucial experiment can play a rhetorical as well as an operational role in science. It does not show that, properly understood, it plays no operational role.[8]

But what is the "proper" operational understanding of crucial experiments? Here there are as many answers as there are approaches to the general problem of theory evaluation in science. For example, on a purely *deductive* account of scientific reasoning, the "logic" of crucial experiments is just the logic of disjunctive syllogism plus modus tollens. Suppose $T_1$ and $T_2$ form an exclusive and exhaustive disjunction. And suppose that $T_1$ implies **O** while $T_2$ implies **Not-O**. The "crucial experiment" yields the "observation" **O**. By modus tollens, $T_2$ is falsified, thus justifying $T_1$ by disjunctive syllogism.

Again, on a *probabilistic* account of theory evaluation, we suppose that $T_1$ and $T_2$ have comparable initial probabilities. A "crucial experiment" would be one for which the final probability of $T_1$ given the observed outcome is much greater than the final probability of $T_2$. Recently it has been argued that theory evaluation is primarily a matter of the relative *explanatory coherence* of the rival theories with given observations. Here the explanatory coherence of a theory is a function of coherence relationships among statements. There is no need here to rehearse the many reasons that might be given for rejecting these approaches to scientific reasoning.[9] I wish only to point out that they all assume a *propositional* account of scientific theories. It is statements that are falsified, assigned low probabilities, or cohere. There is therefore no way that visual or other nonpropositional forms of information can play a role in the reasoning without first being reduced to propositional form. I will now outline an account of crucial experiments that allows reasoning based directly on visual images. It assumes a model-based understanding of scientific theories along the lines outlined above.

The label "crucial decisions" already indicates that my account

|  | $M_1$ | $M_2$ |
|---|---|---|
| Choose $M_1$ |  |  |
| Choose M2 |  |  |

7.2  Decision matrix for a choice between alternative models of the same real system.

of crucial experiments will be formulated within a general account of human judgment. In developing an account of human judgment, one faces a number of alternatives. One is between an account of judgments by individuals or by groups. I shall focus on individuals.[10] Another alternative is whether the individuals in question are to be regarded as "rational agents" or simply scientists. I shall focus on scientists who are idealized only in the sense that the objects of any theory (e.g., classical mechanics) are idealized, not in the sense of providing normative standards.[11]

On anyone's account, a crucial experiment is designed to decide between two well-defined alternatives. The alternatives may be highly specific hypotheses or more broadly conceived "approaches" to the same subject matter. The restriction to two alternatives is not as severe as it might seem. Although, in principle, there are always infinitely many logically possible alternatives, in practice scientists rarely face more than a few. And if there happen to be more than two, they can be dealt with in sequence, two at a time.[12]

So the model of crucial decisions to be employed here is a model of an individual scientist trying to decide between two alternative models, which for the moment we will designate simply as $M_1$ and $M_2$. This yields the standard two-by-two decision matrix shown in figure 7.2. We need only be a little careful how we understand the alternatives. The label "$M_1$," for what decision theorists call "a possible state of the world" should be understood as referring to the possibility that the world is more or less like the idealized model referred to as "$M_1$," or at least that the world is more like $M_1$ than $M_2$. And conversely for $M_2$. The label "Choose $M_1$" means that the

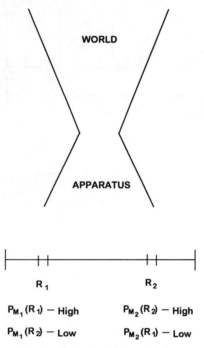

7.3 A schematic representation of a crucial experiment.

agent chooses to regard $M_1$ as providing a satisfactory representation of the world, or at least a better representation than that provided by $M_2$. And conversely for $M_2$.

For present purposes we can take an experiment to be a physical process that yields a reading within a specified one-dimensional range of possible readings, as shown in figure 7.3. What makes an experiment *crucial* are the following conditions:

(i) If the actual world is like model $M_1$, then the experiment is very likely to yield a reading in the range $R_1$ and very unlikely to yield a reading in the range $R_2$.
(ii) If the actual world is like the model $M_2$, then the experiment is very likely to yield a reading in the range $R_2$ and very unlikely to yield a reading in the range $R_1$.

The connection between these conditions and the decision matrix is made by the following obvious "decision rule":

(a) If the experiment yields a reading in the range $R_1$, choose $M_1$.
(b) If the experiment yields a reading in the range $R_2$, choose $M_2$.
(c) If the experiment yields some other reading, reconsider the whole problem.

That this is the appropriate decision rule can be seen simply by running through the possibilities. If the world really is captured by $M_1$, then, by condition (i), the experiment will most likely yield a result in range $R_1$, and, following the decision rule (a), one will choose $M_1$ as providing the better representation of the world. This is clearly the appropriate choice. Similarly, if the world really is captured by $M_2$, then, by condition (ii), the experiment will most likely yield a result in range $R_2$, and, following the decision rule (b), one will choose $M_2$ as providing the better representation of the world. This is again clearly the appropriate choice. Either way, one is very likely to make the "right" choice. Of course, if the reading is something else, there are lots of possibilities, including that neither $M_1$ nor $M_2$ is a very good representation of the world, that the experiment was badly done, etc.

There are many things remaining to be said about this understanding of crucial experiments, and many things said that might be disputed.[13] For present purposes only one aspect requires further clarification. The expressions "very likely" and "very unlikely" in the two conditions stated above must refer to *physical probabilities* (propensities) in the world. They cannot refer to degrees of belief or epistemic judgments. That would lead to a completely different account of crucial experiments. In a single spin of a fair roulette wheel, for example, it is very unlikely that the result will be double zero. It is not just that people attach a low degree of belief to this outcome. Rather, given the physical construction and operation of a roulette wheel, a double zero is physically unlikely. The reason most people give little credence to a belief in this outcome being realized is that they know it is physically unlikely. This is not to say that access to knowledge of physical probabilities is mysteriously direct. On the contrary. These judgments, like all other judgments about the world, are based on more or less definite models of the world. So judgments about physical probabilities, like all judgments about the physical world, are *model-based*.

We are now, finally, ready to proceed to the main objective of this chapter, which is to show how visual presentations of both models and data can be used in crucial decisions about which models best represent the real world.

## 5. Images and Arguments

At this point I will narrow the discussion to the example of twentieth-century geology.[14] Here the alternative scientific theories are better thought of as broadly conceived *approaches to,* or *perspectives on,* geophysics. One approach, commonly labeled *stabilism,* is that the major geological features of the Earth, particularly oceans and continents, originally formed in roughly their current configuration and have remained stable in those positions throughout geological time. The overall mechanism was taken to be cooling, contraction, and solidification of an originally molten sphere. The alternative perspective, *mobilism,* is that the relative positions of the continents and oceans have altered in major ways in geological time, that is, since the original formation of solid land masses. It is a standard part of mobilism, for example, that the Atlantic Ocean is a relatively recent product of a separation of North and South America from Europe and Africa respectively.

During the 1920s, the mobilist cause was championed by Alfred Wegener, a German scientist whose earlier work was in meteorology and atmospheric physics. Wegener provided mobilism with many dramatic visual presentations, most notably a series of three world maps picturing the breakup of Gondwanaland, Wegener's original super-continent containing most of the world's land mass (fig. 7.4). These pictures, which first appeared in the third (1922) German edition of his book, *Die Entstehung der Kontinente und Ozeane,* show the breakup taking place between the Carboniferous (300 million years ago) and the Early Quaternary (500 thousand years ago). I do not claim that these maps constitute the entire content of Wegener's mobilism. Rather, they are visual models which are part of a diverse family of models which all together constitute Wegener's theoretical resources for presenting a mobilist history of the Earth.

VISUAL MODELS | 129

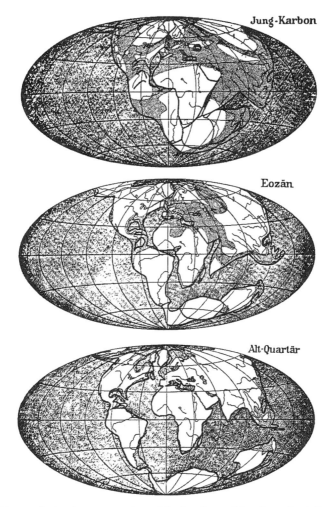

7.4  Wegener's visual representation of the breakup of Gondwanaland. Reproduced from Wegener (1922).

Wegener gathered evidence for mobilism from many domains, including geology, geophysics, paleontology, paleobotany, and paleoclimatology. Here I will concentrate on just one piece of evidence, the celebrated "fit" between the eastern coastlines of North and South America and the western coastlines of Europe and Africa. Figure 7.5 reproduces Wegener's sketch of this fit as it appeared in

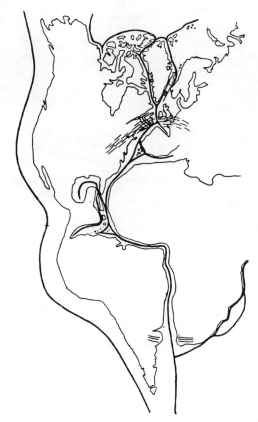

7.5  Wegener's sketch of the Atlantic coastlines indicating continuous mountain ranges across the assumed line of separation. Reproduced from Wegener (1915).

the first (1915) edition of his book. It is a crude sketch. There exist far better drawings exhibiting a better fit dating from over half a century earlier.[15]

What is notable in this sketch is the explicit attention paid to geological features *other* than the fit of the coastlines. In particular, Wegener has marked areas crossing roughly between England and New England, and between South Africa and southern South America, where mountain ranges appear roughly continuous across the postulated border. These congruencies play a major role in his presentation, and apparently did so in his own thinking as well.[16]

Referring to the match in coastlines, Wegener at one point

remarks that it reminds him "of the use of a visiting card torn into two for future recognition" (1924, 44). This is a highly visual metaphor. A little later he modifies and expands the metaphor. Referring to both the match in coastlines and the match in features across the boundary, he writes (1924, 56):

> It is just as if we put together the pieces of a torn newspaper by their ragged edges, and then ascertained if the lines of print ran evenly across. If they do, obviously there is no course but to conclude that the pieces were once actually attached in this way. If but a single line rendered a control possible, we should have already shown the great possibility of the correctness of our combination. But if we have n rows, then this probability is raised to the $n$th power.

Here Wegener appears to go beyond my analysis of crucial decisions, claiming a high probability for his theory itself. But his analysis of the evidence includes the two conditions required for my analysis to be operative. That is, if his mobilist account is correct, then the existence of congruencies like those noted is highly probable. Conversely, if stabilism is correct, and the continents formed independently of one another, then such congruencies are highly unlikely. Since Wegener clearly believes that the congruencies do exist, my analysis could explain why he thinks that mobilism is the obvious choice.

In presenting Wegener's argument I have employed images he himself utilized. What exactly is the role of the images in the presentation? One cannot argue, I think, that the images are *logically* essential. Any information that can be presented in a two-dimensional image can also be presented in a linear, symbolic form, as the digital encoding of images makes obvious. But we are not here concerned with how logically possible scientists might reason. We are concerned with how actual scientists do reason.[17] Wegener's presentation makes it clear both that the images played a large role in his own thinking and that he expected them to play a role in the thinking of his audience as well. But what is that role?

The images, I suggest, function as partial visual models of the relevant features of the Earth. As such they provide grounds for model-based judgments about the physical probabilities that would be operative in the world if it were structured according to the model. Thus, the images provide a basis for the model-based judgments regarding physical probabilities needed in my account of crucial decisions.

Of course, not all the information necessary for making the required probability judgments is present in the image itself. In Wegener's case, for example, one must know that mountain ranges are relatively rare. They do not exist all up and down the coasts of Europe, Africa, and the Americas. Moreover, mountain ranges have distinctive characteristics. So finding several mountain ranges that are congruent across the boundary when the coastlines are lined up according to their matching shorelines is indeed physically unlikely if those mountain ranges had been formed independently on continents separated by thousands of miles.

What happens in such cases, I suggest, is that the visual model serves as an organizing template for whatever other potentially relevant information the agent may possess, regardless of how that information is encoded. The visual image guides the agent's recall of stored information by providing a guide to what, within the agent's diverse store of information, is most relevant to the required probability judgments.

For the purposes of my argument here, it is not necessary that my suggestions regarding the role of images be scientifically correct. It would be nice, of course, if something like this were indeed the case. And it is in line with some current thinking in the cognitive sciences. But all my argument requires is that some such account be physically (and psychologically) possible. That shows at least that images *could* play a significant role in scientific reasoning. And that is enough to refute in principle claims that no such role is possible.

Before moving on to later developments, I would like to illustrate my position with one further image that played a role in the debates over mobilism in the 1920s. Two years after the 1924 publication of the English translation of Wegener's book, *The Origin of Continents and Oceans,* the American Association of Petroleum Geologists sponsored a symposium in New York City on mobilism (van der Gracht 1928). The symposium featured leading scientists

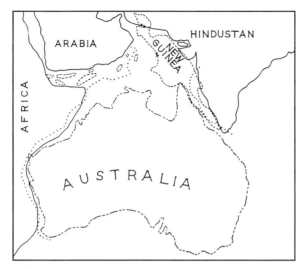

7.6  Longwell's map showing Australia fitting into the Arabian Sea. Reproduced from van der Gracht (1928, 154).

from around the world, including Wegener himself. Of the fourteen participants, roughly one third supported mobilism, one third were genuinely open minded, and one third were strongly opposed to mobilism.

Among the arguments against Wegener in particular was one offered by Yale geologist Chester Longwell based on the map shown in figure 7.6. This map shows a fairly good fit of the coastlines of Australia and New Guinea within that of the Arabian Sea. But no one present, including Wegener, wished to argue that Australia once filled the Arabian sea.

On my analysis of crucial decisions, the real point of this image is to undermine one of the conditions for Wegener's view of the decision in favor of mobilism. Wegener's position requires that it be physically improbable that there be such matching coastlines within a stabilist model of the Earth. Longwell's map provides a clear visual presentation of just such a match. One need not invoke much additional information to be led toward the conclusion that, even within a stabilist model, such matches may not be so improbable as Wegener's position requires. As Longwell himself put it (van der Gracht 1928, 153):

> This case is worth some study, in connection with the better known case of South America and Africa, in order to convince ourselves that apparent coincidence of widely separated coast lines is probably accidental wherever found and should not influence anyone unduly in considering the displacement hypothesis.

In short, there may be visual force on both sides of an argument.

## 6. The Visual Development of Theoretical Models

Wegener died tragically in 1930 on an expedition to Greenland in search of new evidence for mobilism. About the same time, an English geologist, Arthur Holmes, suggested a new mechanism for mobilism. Inspired by the discovery of natural radioactivity, Holmes reasoned that such radioactivity in the Earth might be able to produce sufficient heat to create convection currents of molten minerals just below the Earth's crust. These currents could split the crust and move it laterally great distances before turning downward toward the core. Figure 7.7 reproduces Holmes's visual rendition of this model in which a continental block is ripped in two, creating a new ocean where once land had been.[18]

In spite of its visual power, Holmes's model seems to have done little to stave off the general decline of interest in mobilism following Wegener's death. A good part of the explanation for its lack of immediate influence, I would argue, is that this model provides no basis for a crucial decision between mobilism and stabilism. The processes pictured in Holmes's model would be taking place well below the Earth's crust, too remote for any instruments then known. And contemporary surface manifestations, if any, would take place too slowly to measure. Holmes himself did not even conjecture a possible crucial experiment.

Holmes's convection model was revived thirty years later by the American geologist Harry Hess. Figure 7.8 reproduces Hess's version of the model (Hess 1962, 607). The main difference is that Hess has the convection current rising under the ocean floor rather than a continental block. This was because Hess, unlike Holmes, intended his model to explain the origin of the great ocean ridge systems first

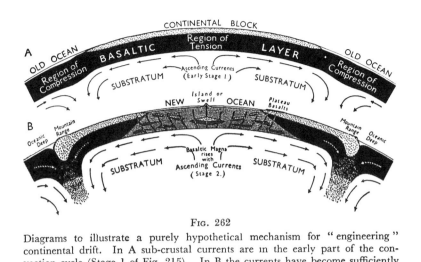

FIG. 262
Diagrams to illustrate a purely hypothetical mechanism for "engineering" continental drift. In A sub-crustal currents are in the early part of the convection cycle (Stage 1 of Fig. 215). In B the currents have become sufficiently vigorous (Stage 2 of Fig. 215) to drag the two halves of the original continent apart, with consequent mountain building in front where the currents are descending, and ocean floor development on the site of the gap, where the currents are ascending

7.7  Holmes's dynamic visual representation of convection currents splitting a continent to produce a new ocean. Reproduced from Holmes (1944, 506).

explored in the 1950s. The ridges, on Hess's model, are produced directly above the rising convection current which then spreads out, creating a new sea floor. But Hess's model, like Holmes's, provides no basis for a crucial decision between mobilism and stabilism.

The makings of a crucial experiment were provided by a new graduate student in geophysics at Cambridge, Fred Vine, and his recently appointed supervisor, Drummond Matthews. In late 1962, Matthews returned from an expedition to the Indian Ocean where he had obtained systematic measurements of the total magnetic field at the level of the ocean floor across the Carlsberg Ridge. The task of analyzing this magnetic data fell to Vine, while Matthews went off on his honeymoon.

Using then very new computer techniques, Vine determined that the magnetic readings across the ridge showed a small periodic

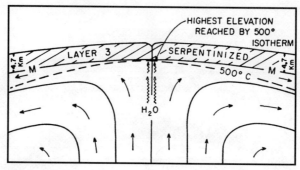

7.8   Hess's dynamic visual model of sea floor spreading produced by convection currents. Reprinted by permission from Hess (1962, 607).

variation as one moved away from the center of the ridge. Similar periodic variations in magnetic intensity along the ocean floor had earlier been observed in other areas, such as the Pacific Ocean off the coast of North America. Taking these variations in magnetic intensity as a real phenomenon requiring explanation, Vine, early in 1963, set about finding one.

Vine was keenly aware of Hess's model of sea-floor spreading, having seen Hess himself present it during a conference at Cambridge in January, 1962. A possible link to the observed variations in magnetic intensity was provided by initially unrelated work on paleomagnetism. Researchers in California, led by Allan Cox, had been examining the direction of remanent magnetism in core samples from lava flows. Such samples provide a measure of the direction of the Earth's magnetic field when the examined material was molten, since magnetic material in a molten fluid would tend to line up with the existing magnetic field of the Earth. What they found, confirming scattered findings dating to half a century earlier, was an apparent change in direction of the Earth's magnetic field several times in the past four million years.

Figure 7.9 shows a visual presentation of both the data and the theory in one of the first publications, in mid-1963, of the California group (Cox, Doell, and Dalrymple 1963). The single vertical scale, representing age in millions of years, starts with zero at the top and shows increasing time into the past as one moves down the scale. This arrangement represents the obvious geological fact that in a lava flow the younger materials from recent eruptions are toward the top while the older materials are deeper. Each data point represents a number of rock samples from a given site, with the average age of the samples indicated by the location of the data point relative to the vertical scale. The polarity of the sample, "normal" or "reversed," is indicated by its location in the left or right column. The rival models are simple. They just represent the magnetic field of the Earth as having been continuously normal for a time into the past, then being reversed, then being normal, and so on in equal time intervals. One model puts the period of the reversals at a half million years, the other at a million. That both models are consistent with the data can be seen immediately in the graphical presentation.

What is the connection between (i) Vine's data showing regular variation in magnetic field intensity extending out from an ocean ridge, (ii) Hess's model of sea-floor spreading, and (iii) evidence for geomagnetic reversals? The answer cries out for a dynamic, visual model, but none seem to have been published during the crucial years, 1963–66. Vine and Matthews's 1963 paper contains, instead, the following verbal description (Vine and Matthews 1963, 948):

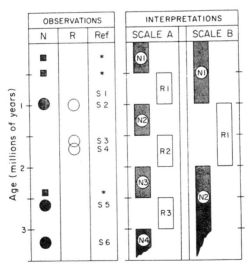

7.9 The first visual presentation of data and models by the California group investigating geomagnetic reversals. Reprinted with permission from *Nature* 198:1049–51. Copyright 1963 Macmillan Magazines Limited.

> The theory is consistent with, in fact virtually a corollary of, current ideas on ocean floor spreading and periodic reversals in the Earth's magnetic field. If the main crustal layer . . . of the oceanic crust is formed over a convective up-current in the mantle at the centre of an oceanic ridge, it will be magnetized in the current direction of the Earth's field. . . . Thus, if spreading of the ocean floor occurs, blocks of alternately normal and reversely magnetized material would drift away from the centre of the ridge and parallel to the crest of it.

No one can deny, however, that in reading this description it helps to refer back to the dynamic visual models of Holmes and Hess.[19]

Following the above description is a visual presentation of the magnetic data and a corresponding model of the sea floor across three different ridges (fig. 7.10). This appears to be an adaptation of Cox's model for geomagnetic reversals, except that the blocks of alternately magnetized material are laid out horizontally rather than vertically. This difference, of course, reflects the differing causal processes suggested as having produced the two configurations of differentially magnetized materials.

Publication of Vine and Matthews's paper seems to have convinced almost no one of the reality of sea floor spreading and the mobilism it implies. Not even they were willing to claim they had proven the case. In the last few lines of their 1963 article they write (Vine and Matthews 1963):

> It is appreciated that magnetic contrasts within the oceanic crust can be explained without postulating reversals of the Earth's magnetic field; for example, the crust might contain blocks of very strongly magnetized material adjacent to blocks of material weakly magnetized in the same direction. However, the model suggested in this article seems to be more plausible because high susceptibility contrasts between adjacent blocks can be explained without recourse to major inhomogeneities of rock type within the main crustal layer or to unusually strongly magnetized rocks.

In terms of my model of crucial decisions, they do seem to think that condition (i) is satisfied. The results obtained are fairly probable given a model incorporating sea floor spreading and geomagnetic reversals. But these results are not wildly improbable if those assump-

# VISUAL MODELS | 139

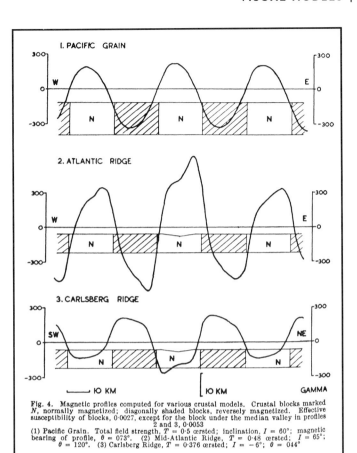

7.10   Vine and Matthews models for magnetic profiles near ocean ridges. Reprinted with permission from *Nature* 199:947–49. Copyright 1963 Macmillan Magazines Limited.

tions are mistaken and stabilism is correct. So there is no adequate basis for making a crucial decision in favor of sea floor spreading and mobilism.

Of course the noted possibilities for stabilist explanations of the data are not directly contained in the visual presentation of their model of the sea floor or of their data. But realizing that the simple periodic structure of the model was just read off the similarly simple periodic structure of the data makes it easy visually to assimilate suggested alternative models. So the visual presentation facilitated

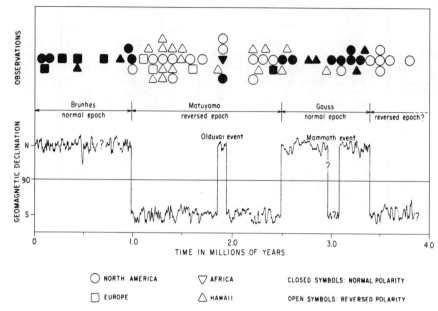

7.11  The second visual presentation of data and models by the California group investigating geomagnetic reversals. Reproduced with permission from A. Cox, R. R. Doell, and G. B. Dalrymple, Reversals of the Earth's Magnetic Field. *Science* 144:1537–43. Copyright 1964, AAAS.

the judgment that the prospects for a crucial decision were not yet compelling.

During the next two years the conditions for a crucial decision improved in one respect, but declined in another. In 1964, the California group (Cox, Doell, and Dalrymple 1964), having acquired data from several new sites, published a new scale of geomagnetic reversals (fig. 7.11). Two major differences from the earlier scale are immediately evident. First, they have given up the assumption of equal time periods of normal and reversed polarity. The major normal and reversed "epochs" are now of irregular duration. Second, they have refined the scale to include several brief (100 thousand year) "events." The Olduvai event, around 1.9 million years ago, is a brief period of normal magnetism within a long epoch of reversed magnetism. Similarly, the Mammoth event, around 3 million years ago, is a brief period of reversed magnetism within an epoch of normal magnetism.

Meanwhile, Vine and a visiting senior Canadian geologist, Tuzo Wilson, were busy analyzing magnetic data from yet another ridge system, this one in the Pacific off the coast of Vancouver. Figure 7.12 reproduces one of their visuals as published in 1965 (Vine and Wilson 1965). Following the California group, their models now exhibit both unequal epochs and several briefer events. In terms of my own model for crucial decisions, the good news was that an aperiodic pattern of reversals with intervening small events is very unlikely to appear in scattered places around a stabilist Earth. On any stabilist model, it would take a near miracle for the possible sources of magnetic variation in the sea floor near ridges to produce the same complicated irregular pattern near several different ridges lying in different oceans.

The bad news was that the pattern they were finding in the magnetic sea floor data was not exactly what the Cox scale would lead one to expect if the Vine-Matthews model were correct. If one holds to the constraint that the spreading rate of the sea floor has been roughly constant, there was no way to match up the observed magnetic readings across the ridge with the new scale of reversals published by the California group. The various epochs and events simply did not line up as expected. What was gained in the satisfaction of one condition was lost in failure to satisfy the other.

## 7. The Persuasive Power of Images

Within a year the situation had changed dramatically. In late 1965 a research vessel operated by the Lamont Geological Observatory of Columbia University returned from a new geological survey of the Pacific-Antarctic Ridge with the dramatic magnetic profile shown in figure 7.13 (Pitman and Heirtzler 1966, 1166). Whereas earlier profiles and geomagnetic time scales had extended out to around four million years ago, this profile extended out a distance corresponding to ten million years, revealing a continuing pattern of reversals never before detected.

The bilateral symmetry of the profile is of particular significance, as is the method used to make it visually obvious. The center profile shows the magnetic readings moving from west to east at the right of

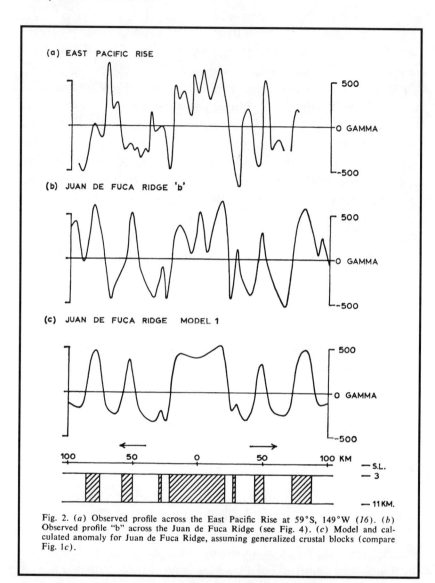

Fig. 2. (a) Observed profile across the East Pacific Rise at 59°S, 149°W (*16*). (b) Observed profile "b" across the Juan de Fuca Ridge (see Fig. 4). (c) Model and calculated anomaly for Juan de Fuca Ridge, assuming generalized crustal blocks (compare Fig. 1c).

7.12  Vine and Wilson's visual comparison of magnetic data and crustal model for the Juan de Fuca Ridge. Reproduced with permission from F. J. Vine and J. T. Wilson, Magnetic Anomalies over a Young Oceanic Ridge off Vancouver Island. *Science* 150:485–89. Copyright 1965, AAAS.

VISUAL MODELS | 143

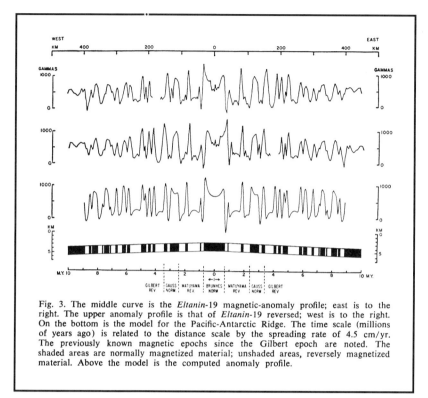

Fig. 3. The middle curve is the *Eltanin*-19 magnetic-anomaly profile; east is to the right. The upper anomaly profile is that of *Eltanin*-19 reversed; west is to the right. On the bottom is the model for the Pacific-Antarctic Ridge. The time scale (millions of years ago) is related to the distance scale by the spreading rate of 4.5 cm/yr. The previously known magnetic epochs since the Gilbert epoch are noted. The shaded areas are normally magnetized material; unshaded areas, reversely magnetized material. Above the model is the computed anomaly profile.

7.13  A visual comparison of magnetic profiles of the Pacific-Antarctic Ridge with a corresponding model. Note especially the symmetry described in the text. Reproduced with permission from W. C. Pitman and J. P. Heirtzler, Magnetic Anomalies over the Pacific-Antarctic Ridge. *Science* 154:1164–71. Copyright 1966, AAAS.

the diagram. The top profile is just the middle profile reversed, with west on the right of the diagram. Merely by scanning visually across the diagram and comparing these two profiles one can see just how amazingly symmetrical the profile is. That it should be symmetric is an immediate consequence of the Vine-Matthews model, since the sea floor should spread out equally on both sides of a ridge. The lower profile is derived from the model shown at the bottom of the diagram.

About the same time as the data from the Pacific-Antarctic

144 | CHAPTER SEVEN

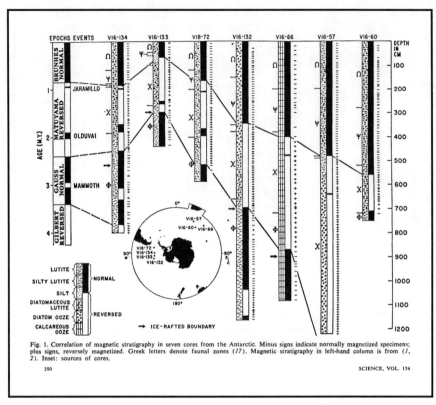

7.14  A visual comparison of magnetic reversals in deep-sea sediments with the time scale for reversals in terrestrial lava flows. Reproduced with permission from N. D. Opdyke, B. P. Glass, J. D. Hays, and J. H. Foster, Paleomagnetic Study of Antarctic Deep-Sea Cores. *Science* 154:349–57. Copyright 1966, AAAS.

Ridge were being analyzed, another group at Lamont was busy analyzing the magnetic orientation of sedimentary materials in core samples taken from the ocean floor near the tip of South America. Like cooling lava, sediment traps magnetic materials in their existing spatial orientation as the sediment packs more tightly. If the pattern of geomagnetic reversals is real, it should also be recorded in such sediments. Again the evidence, as presented visually in figure 7.14, is dramatic (Opdyke et al 1966, 350). Just by inspecting the diagram, one can see almost immediately that the pattern formed by regions of normal and reversed magnetism within the core samples closely matches that of the magnetic profiles across ocean ridges.

VISUAL MODELS | 145

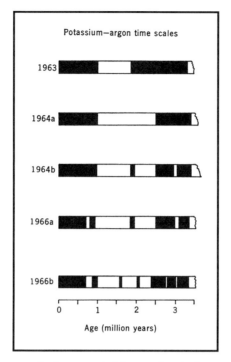

7.15  Cox's summary of refinements in his group's scale for geomagnetic reversals during the years 1963–66. Reproduced with permission from A. Cox, Geomagnetic Reversals, *Science* 163:237–45. Copyright 1969, AAAS.

But what of the mismatch between the geomagnetic times scales of the California group and the sea floor profiles which had plagued Vine and Wilson just one year earlier? That too was resolved. Working with rock samples discovered near Jaramillo Creek in New Mexico, several members of the California group discovered that the current period of normal magnetism extended not one million years into the past, but only about .7 million years. It was followed by a brief period of reversed magnetism and then an event of normal magnetism extending between .9 and 1.0 million years ago. Vine and Wilson had gone astray because they had identified the first normal event in their profiles as the nearly two million year old Olduvai event rather than the one million year old Jaramillo event. Figure 7.15 presents Cox's retrospective summary of his group's results during the crucial years 1963–66 (Cox 1969, 239).

Versions of these last three images were all presented in a historic session of the April 1966 meeting of the American Geophysical Union. By all accounts the effect was dramatic. And so it should have been if my account of crucial decisions is correct. That the data obtained were to be expected if the Vine-Matthews model is correct was well known. What made these presentations especially dramatic was that they showed how utterly improbable the data would be on any stabilist model. What stabilist process (short of divine creation) could possibly have produced that visually dramatic and detailed signature pattern simultaneously in widely scattered continental lava flows, deep sea sediments, and the floor of several different oceans? This was visually obvious to all, regardless of their particular research specialties. As Allan Cox, who chaired the April AGU session, later summed it up, ". . . there was just no question any more that the seafloor-spreading idea was right" (Glen 1982, 339).

## 8. Conclusion

This chapter connects several recent themes in the philosophy of science, and in science studies more generally: (i) a model-based picture of scientific theories, (ii) a naturalistic account of crucial decisions, and (iii) interest in the use of visual images in scientific thinking. To structure the presentation, I assumed a model-based account of scientific theories and used the fact that it could accommodate visual information to support my naturalistic account of crucial decisions. But the argument need not have been structured this way. Probably the most appropriate conclusion is that these three themes are mutually reinforcing, and together support a move away from an exclusive reliance on propositional modes of analysis for understanding the workings of modern science.

# PART
## THREE

Perspectives on the Philosophy of Science

## Introduction

The essays in part 3 were originally intended for an audience consisting primarily of my fellow philosophers of science. For others, and even for newcomers to the philosophy of science itself, some orientation is in order. This applies particularly to the first two essays, "Philosophy of Science Naturalized" and "Constructive Realism."

In the social sciences, including the sociology of science, *naturalization* typically refers to an illicit process of reading social arrangements into the natural order of things, particularly the biological order. In philosophy the intent is different. To naturalize a part of philosophy is merely to treat it as an empirical subject with the same epistemological status as any natural science. In practice, this means eliminating any reliance on synthetic *a priori,* or otherwise transcendental, claims. Within the general science studies community, then, virtually everyone except some philosophers is a naturalist in this latter sense. Thus, it tends to be only among philosophers of science that naturalization is an issue.

Both of the above-mentioned essays refer to a doctrine called "constructive realism." Here one might well inquire after the relationship, if any, between this doctrine and constructivism in the sociology of science. And how, one might ask, is constructive realism related to my own notion of "perspectival realism" discussed in earlier chapters?

Constructive *realism* was originally intended as an alternative to the constructive *empiricism* of the philosopher of science, Bas van Fraassen. Van Fraassen's view is that scientific theories are *logical,* not social, constructs. His view is *empiricist* in that he holds that the only epistemological constraint on theories is that they correctly account for *observable* aspects of reality. The practice of science, according to van Fraassen, licenses no claims about non-observable aspects of reality. I agree with van Fraassen about theories being logical constructs, but disagree that the legitimate claims of scientists are restricted to the merely observable realm. I argue that scientists legitimately make claims also about the hidden causal structure of the world. Thus, constructive *realism.* Only later did I realize that the term "constructive realism" is also nicely opposed to constructivism in the sociology of science. It suggests agreement that there is indeed something importantly constructive about doing science. Science is not merely the discovery of already categorized objects and relations. The categories we use are to some extent constructed by us. Nevertheless, scientists can sometimes legitimately claim genuine similarities between their logical constructs and aspects of reality. That makes me some kind of realist rather than a social constructivist.

*Perspectival Realism,* finally, is a later development of constructive realism. The constructive element remains as before. The difference is the insistence that our theories do not ever capture the totality of reality, but provide us only with perspectives on limited aspects of reality. Scientific knowledge is not absolute, but perspectival.

# EIGHT

# Philosophy of Science Naturalized

## 1. Kuhn's Naturalism

In the very first chapter of *The Structure of Scientific Revolutions,* Kuhn (1962) sought to establish "a role for history." Part of that role, he implied, is as data for "a theory of scientific inquiry." And by "theory" he meant something comparable to theories in the sciences themselves. Thus, referring to standard philosophical distinctions, such as that between "discovery" and "justification," he wrote (1962, 9):

> Rather than being elementary logical or methodological distinctions, which would thus be prior to the analysis of scientific knowledge, they now seem integral parts of a traditional set of substantive answers to the very questions upon which they have been deployed. That circularity does not at all invalidate them. But it does make them parts of a theory and, by doing so, subjects them to the same scrutiny regularly applied to theories in other fields. If they are to have more than pure abstraction as their content, then that content must be discovered by observing them in application to the data they are meant to elucidate. How could history of science fail to be a source of phenomena to which theories about knowledge may legitimately be asked to apply?

Although he did not use exactly these words, Kuhn was advocating a *naturalized* philosophy of science.

The many philosophical criticisms of Kuhn's work focused mainly on the details of his naturalistic account. His account of revolutionary theory change, which invokes only naturalistic notions like gestalt switches and persuasion, was a frequent target. Few critics, however, raised the general question whether *any* purely naturalistic theory of science might be correct. No doubt this was due to the unquestioned presumption that no such account could be correct. It is precisely this presumption I wish explicitly to challenge.

For some, I admit, no challenge is necessary. Some philosophers regard the philosophy of science as merely a branch of epistemology. And some of these philosophers follow Quine (1969) in the project of naturalizing epistemology. Others embrace a version of evolutionary epistemology. But these are still minorities. Naturalism in the philosophy of science is generally rejected not only by the successors to logical empiricism but also by most of those who agree with Kuhn in adopting a historical methodology. Thus Lakatos (1970), Toulmin (1972), Laudan (1977), and Shapere (1984) have each sought to show that the process of scientific inquiry is not only historical but *rational* as well. Rationality is not a concept that can appear in a naturalistic theory of science—unless reduced to naturalistic terms.[1]

My arguments will be both negative and positive. I will first exhibit what seem to be the main general objections to a naturalistic approach. These objections, I suggest, are too strong. There seems no viable non-naturalistic way around them either. I review the failure of two such non-naturalistic approaches, methodological foundationism (Carnap, Reichenbach, and Popper) and metamethodology (Lakatos and Laudan). The correct response, I suggest, is to adopt an "evolutionary perspective." This perspective is defended against one critic, Hilary Putnam. To argue the plausibility of a naturalistic approach, I next sketch a naturalistic account of theories and of theory choice. In conclusion, I return to Kuhn's question about the role of history in developing a naturalistic theory of science.

## 2. Some Arguments against Naturalism

The following are some general forms of argument that one would expect to be raised against any proposal to naturalize the philosophy of science.

*The Circle Argument*

The general idea behind the circle argument is that the use of scientific methods to investigate scientific methods must be circular, beg the question, or lead to a regress. A more explicit version of the argument might go something like this: One of the things any study of science must investigate is the methods (criteria, canons, etc.) scientists use in evaluating evidence. To pursue such an investigation *scientifically* requires using data about scientific practice to reach conclusions about scientific methods. Thus, any empirical investigation aimed at discovering the criteria that scientists use for evaluating evidence would necessarily presuppose at least some of the criteria it was supposedly setting out to discover. So not *all* the methods of science could be discovered by scientific investigation. At least *some* must be discoverable by other means.[2]

The circle argument is a version of classic arguments concerning the justification of induction. This relationship may partly explain the power of the argument within the philosophical community.

*The Argument From Norms*

This argument appeals to the distinction between facts and norms. A naturalistic study of science, it is claimed, could at most *describe* the methods scientists use in coming to adopt hypotheses or theories. The goal of the philosophy of science, however, is not merely to describe the methods scientists employ, but to *prescribe* what methods they *should* employ. We want to know not merely what criteria scientists in fact use in adopting theories; we want to know which are the *right* criteria. A naturalistic philosophy of science would be powerless to answer such questions.

*The Argument From Relativism*

This argument may be viewed as a corollary to the argument from norms, but relativism has been so much discussed of late that this form of argument deserves independent billing. The argument has the form of a *reductio*. A naturalistic philosophy of science, it is claimed, would be powerless to distinguish good from bad science. It would, for example, have to treat "creation theory" on a par with evolutionary theory. Such a philosophy of science would be at best worthless, at worst pernicious.

These, in brief, are some of the main arguments against a naturalized philosophy of science. A reply takes a bit longer.

## 3. Methodological Foundationism

Circle arguments have always been among the most powerful in the philosopher's arsenal. Their use, however, commits the user to constructing a defense against a similar attack. Regarding our particular circle argument, the traditional defense has been some form of *methodological foundationism*—the construction of a method whose correctness can be certified *a priori*.

One connection between naturalism and foundationism has been well charted by Quine (1969). For Quine it was the foundationist inability to reduce mathematics to logic (or semantics or behavior) that left us no alternative but to naturalize epistemology. It requires only a little elaboration to see that a similar connection exists within the philosophy of science.

When Kuhn's book first appeared, the methodologically foundationist programs of Carnap, Reichenbach, and Popper were among the most active areas of research in the philosophy of science. Carnap and Reichenbach, though not Popper, were also tempted by foundationism with regard to the data furnished by experience. Since this latter aspect of foundationist programs is not at issue here, I shall say no more about it.

Carnap was originally attracted to Russell's foundationism which utilized the method of logical construction developed in the context of the foundations of mathematics. Here the regress stops at a foundation of *logic*. Logic also provides a normative component for scientific reasoning and a bulwark against relativism.

Carnap and the early Logical Empiricists gave up Russell's strict form of methodological foundationism not so much for technical reasons as for broadly empirical reasons. They concluded that the methods of logical construction could not yield the laws of physics as they understood them, and they were unwilling to reject the laws of physics as philosophically unsound. But few were willing to give up the idea that logic provides the foundation for scientific method.[3]

Even Popper, who otherwise was quite critical of many Logical Empiricist doctrines, rested his methodology on the simple logical rule of *modus tollens*. Here again, some of the most severe criticisms of Popper's methodology have been broadly empirical. Adopting Kuhn's claim that all theories have faced anomalies throughout their careers, Lakatos (1971), for example, argued that if we follow Popper's rules, we should have to regard all theories as falsified. Assuming falsified theories should be rejected, we reach the empirically unacceptable conclusion that all theories should be rejected.

Among the major Logical Empiricist figures, Reichenbach was unusual in seeking a methodological foundation not in logic but in a pragmatic rule of action. Assuming that the ultimate goal of science is to determine limiting relative frequencies in infinite sequences, he offered an *a priori* argument that his inductive rule of inference, the "straight rule," guaranteed success in reaching this goal so long as the goal was obtainable at all. Unfortunately, there is a continuum of "crooked rules" for which the same justification can be given. Hacking (1968) delivered the *coup de grâce* to the program by showing that any long-run justification sufficient to justify only the straight rule required the *empirical* assumption that the sequences in question be *random*. This exposed the program to the regress argument it was explicitly designed to elude. Reichenbach's program could also have been criticized on the *empirical* ground that science in fact has stronger goals than the long-run discovery of limiting relative frequencies. But this was not the argument that led to its demise.

During the 1940s, after moving to the United States, Carnap took up Keynes's program of developing an *inductive logic* that would be a formal generalization of deductive logic. His own semantic theories of the previous decade provided the formal background for this attempt. Being a logic, this inductive logic would stop the regress and provide the norms to defeat relativism. Carnap's inductive logic has been criticized on the empirical ground that it is too simple to tell us anything about the evaluation of actual scientific theories. The main reason most people gave up the program, however, was more technical. The logic requires a measure of the initial probability of all hypotheses. But the space of possible measures, like the space of Reichenbachian rules of inference, is so large that there seems no *a priori* and non-arbitrary way to justify a unique measure.

Recalling that the first chapter of Carnap's *Logical Foundations of Probability* (1950) was entitled "On Explication," we are reminded of another program for grounding a particular inductive logic, namely, as an explication of our prereflective concept of evidential support. But how do we know when we have correctly captured this concept, or even that "we" have a univocal concept of this sort? Carnap says only that the "explicatum" must be similar to the "explicandum." If this similarity is to be determined empirically, and there seems no other way, then Carnap too was caught in the circle argument.

Richard Jeffrey (1973) once argued that the correct inductive logic would be the one that eventually agrees with our inductive intuitions when the Carnapian program is sufficiently developed for more complex languages. To avoid the circle, this view must assume that our intuitions are given directly without empirical investigation. Moreover, this interpretation leaves it open whether the program ever will be sufficiently developed—which makes it impossible for us currently to say whether science ever has been or is now a rational enterprise. This is a rather weak foundation. And like any explication, it provides no protection from relativism. The logic is at best *descriptive* of *our* intuitions. It does not insure us that our intuitions themselves are correct.

The major remaining stronghold of methodological foundationism is "Bayesian inference" or one of its near relatives. Here the problem of picking out a unique initial probability measure is avoided by relativizing to an individual agent. Individuals supply their own initial measures. Rationality then consists in how one *revises* probability assignments in light of new evidence. It has often been objected that Bayesian inference goes too far in the direction of relativism. Moreover, the same type of problem that plagued both Carnap and Reichenbach arises here too because there are many different logically possible ways of "conditionalizing" on the evidence, and no *a priori* way of singling out one way as uniquely rational. One is reduced either to appealing to something like "explication," or to investigating actual reasoning, which reintroduces the circle.[4]

Adopting Quine's form of inference, I should like now to conclude that methodological foundationism is a hopeless program and thus that naturalism, in spite of the circle argument, is our only alternative. There is, however, a further line of inquiry that must be considered, if only because it has been so prominent in the post-Kuhnian literature.

## 4. Metamethodology

Imre Lakatos (1971) introduced the term "metamethodology" to describe his method for investigating the relative superiority of any proposed theory of scientific method. Laudan (1977) adopted a similar strategy—though differing in detail. For brevity of exposition, I will concentrate on Laudan's approach.[5]

The connection between metamethodology and the circle argument arises as follows. If Lakatos and Laudan really had been taking a naturalistic approach to methodology, they would have adopted the reflexive strategy of applying their methodology to itself. This, however, is not their official doctrine. Whether they deliberately avoided a reflexive strategy *because* of its obvious circularity, I cannot say. Their metamethodologies, however, are not reflexive and thus not blatantly circular. Whether they can achieve their ends while still avoiding circularity is another question.

In Laudan's theory of scientific rationality, the measure of progress, and therefore of rationality, is problem-solving effectiveness. Roughly speaking, the problem-solving effectiveness of a research tradition is the weighted number of empirical problems solved by its latest theory minus the weighted number of outstanding anomalies and conceptual problems. Problems are weighted by their importance to the research tradition. The relative acceptability of one research tradition over another is determined by its relative problem-solving effectiveness. This measure, Laudan claims, provides a *rational* way of deciding the relative acceptability of two research traditions.

Laudan's metamethodological strategy is to seek first a set of "preferred preanalytic intuitions about scientific rationality" (PIs). That is, looking at the history of science, we find

> a subclass of cases of theory-acceptance and theory rejection about which most scientifically educated persons have strong (and similar) intuitions. This class would include within it many (perhaps even all) of the following: (1) it was rational to accept Newtonian mechanics and to reject Aristotelian mechanics by, say, 1800; . . . (4) it was irrational after 1920 to believe that the chemical atom had no parts; . . . (1977, 160)

The next step is to apply the methodology (Laudan's theory) to the PIs in order to determine the relative problem-solving effectiveness of the traditions in question. This assumes, of course, that we can indeed identify, count, and weigh the relevant problems. Comparison of computed problem-solving effectiveness will tell us which tradition was in fact most progressive and thus which should have been accepted according to Laudan's methodology. "The degree of adequacy of any theory of scientific appraisal is proportional to how many of the PIs it can do justice to" (1977, 161).

Assume, for the sake of argument, that Laudan's methodology agrees with *all* the PIs. Could we then be confident that it is "a sound explication of what we mean by rationality" (1977, 161)? I think not. The most one could conclude is that Laudan has identified a highly reliable *symptom* of the basis for our preanalytic judgments of theory-acceptance and theory-rejection.

Suppose, contrary to Laudan, that our preanalytic judgments are really based on an assessment of the approximate *truth* of the theories in question, and that we take problem-solving effectiveness as our best *evidence* for approximate truth. Laudan's method of assessment would then yield the same judgments of acceptance and rejection, but fail to capture the real basis of our judgments. The trouble is that comparison with our gross judgments of acceptance and rejection does not test the fine structure of the methodological theory. To test the fine structure, however, would require a more detailed empirical inquiry, and this would immediately raise the problem of circularity.

It may be, however, that Laudan would not be all that bothered by learning that he has not avoided circularity. His main concern, like that of Lakatos before him, is to have a *normative* theory of rationality. Let us, therefore, move on to the argument from norms.

Does the type of metamethodology advocated by Lakatos and Laudan yield methodological principles which are genuinely normative? Not really. At its most successful, the metamethodology would tell us only that we had discovered a general *description* of situations which we intuitively regard as clear cases of rational acceptance or pursuit. We might have correctly identified the descriptive component of the methodology, without capturing its normative force. To claim that we had captured the normative component would require that we make the judgments we do *because* of considerations based on problem-solving effectiveness. In Kantian terms, Laudan's metamethodology could at most show only that we are acting in accord with his methodology, not that we are acting out of regard for that methodology. It cannot show that his methodology is actually embodied as a norm in our judgments.

This point is all the more pronounced if we consider not merely our own current preferred intuitions, but those of the historical actors in the episodes considered. Laudan does not attempt to show that actual scientists in historical contexts made the judgments they did because of considerations of problem-solving effectiveness. He is content to point out the correlation between their judgments and our calculations of actual problem-solving effectiveness. That is scant evidence that such considerations were normatively operative at the time.

Being at bottom a strategy for explication, not justification, Laudan's metamethodology also fails to provide a strong defense against relativism. Questions about the rationality of the whole western scientific tradition are ruled out because the metamethodology begins with the assumption that some judgments (the PIs) are rational. It is these we use to test the theory of rationality. This Laudan freely admits (1977, 161). He fails to point out, however, that this leaves us defenseless against the cultural anthropologist who claims that the belief systems of non-Western cultures cannot rationally be judged by the standards of Western science.

## 5. An Evolutionary Perspective

I conclude that neither methodological foundationism nor metamethodology can break the circle or provide the norms needed to defeat relativism. This hardly proves that there is no way to achieve these ends. It does, however, provide some motivation for seeking to *understand* how a naturalized philosophy of science might fruitfully be pursued. I would suggest that evolutionary theory, together with recent work in cognitive science and the neurosciences, provides a basis for such an understanding. The following is the barest sketch of how the story might go.[6]

Human perceptual and other cognitive capacities have evolved along with human bodies. We share many of these capacities with other primates and even lower mammals. Indeed, those parts of our brains responsible for our more advanced linguistic abilities are built upon and linked to those parts that we share with other mammals. There can be no denying that these capacities are fairly well adapted to the environment in which they evolved. Without considerable adaptation, we would very likely not be here. Nor are these capacities trivial. The amount of perceptual and neural processing required just for a human to walk without falling or bumping into things is fantastically large and very complex.

The capacities evolution favors, of course, are just those that confer biological fitness, that is, the ability to survive and leave offspring. The ability to do modern science had nothing to do with the evolution of our perceptual and cognitive capacities—indeed, doing science may very well be detrimental to our survival as a species. The general problem faced by a naturalistic philosophy of science, then, is to explain how creatures with our natural endowments manage to learn so much about the detailed structure of the world—about atoms, stars and nebulae, entropy and genes. This problem calls for a *scientific* explanation.

Empiricist philosophers emphasized the role of immediate perceptual experience in their analyses of knowledge because of the high degree of subjective certainty attached to such experience. From an evolutionary perspective, the subjective certainty is indeed causally connected with the more direct source of the reliability of such judgments, which lies in our evolved capacities for interacting with our world. But the operation of these capacities is largely unrecorded in our conscious experience. Rationalist philosophers, on the other hand, focused on our more general subjective intuitions, such as that space has three dimensions and that time exhibits a linear structure. These judgments seem to be built into the way we think. And indeed they are, for the aspects of the world relevant to our biological fitness have roughly that structure.

Neither empiricists nor rationalists could see how to get beyond their subjective experience or intuitions. This led to the familiar philosophical views that the world is nothing more than the sum total of our sense experience or that it is totally unknowable. In fact, we possess built-in mechanisms for quite direct interaction with aspects of our environment. The operations of these mechanisms largely bypass our conscious experience and linguistic or conceptual abilities.[7]

Thinkers struggling to understand the nature of their own knowledge in the seventeenth and eighteenth centuries may be forgiven for not appreciating evolutionary theory or contemporary neurobiology. A century after Darwin, a similar lack of appreciation is less forgivable.

The traditional philosophical skeptic would of course seek to reintroduce the circle argument. To invoke evolutionary theory to understand how we know about the world, he would say, simply begs the question. Evolutionary theory is a fairly advanced, and therefore problematic, form of scientific knowledge. Our problem is to justify that knowledge using something less problematic, such as what one can "directly" experience or intuit.

At this point, however, the skeptic's reply is equally question-begging. Three hundred years of modern science and over a hundred years of biological investigation have led us to the firm conclusion that no humans have ever faced the world guided only by their own subjectively accessible experience and intuitions. Rather, we now know that our capacities for operating in the world are highly adapted to that world. The skeptic asks us to set all this aside in favor of a project that denies our conclusion. And he does so on the basis of what we claim to be an outmoded and mistaken theory about how knowledge is, in fact, acquired.

It should be noted that the above appeal to evolutionary theory is far more modest than that of numerous advocates of "evolutionary epistemology." It is limited to explaining why we need not worry about our failure to break the circle argument. Others have advanced the more extensive claim that evolutionary theory itself provides a good model for the overall development of scientific knowledge. I doubt that it is a very good model for this more ambitious purpose, and I shall suggest a far different account. We agree, however, that the issue is an empirical one, to be settled by scientific procedures, and not by purely philosophical arguments.[8]

Finally, an evolutionary perspective provides a program for dealing with norms and the problem of relativism. At some stage in the evolutionary process, the evolution of human organisms and that of human societies became coextensive. Even modestly complex societies require some social organization. Norms make it possible to maintain the requisite degree of social organization. Nor need the naturalist regard these as mere regularities in social behavior. Norms are taught and enforced by various means of social control. The regularity is a product of these social actions. What the naturalist denies is that there is any basis for the norms that transcends the society in its actual physical context. But does this view not leave us open to a radical form of relativism?

An evolutionary perspective places definite limits on how different a human society on earth could be. It is not physically possible that there should exist on earth a culture totally alien to us. Humans walk, talk, eat, sleep, and procreate. Correspondingly, they must acquire food and shelter. We could not fail to understand these activities. How a society goes about doing these things, on the other hand, is not uniquely determined by our biological nature, even if we include the physical circumstances of that society. There is always more than one way to skin a cat. Moreover, there is no supracultural basis for the norm that cats are to be skinned one particular way (or perhaps not at all). At this level, cultural relativism is correct. Does this imply that "creation theory" is as good as evolutionary theory? No more than it implies that prayer is as effective as penicillin for curing infections. Vindicating this reply, however, requires a positive theory of science.

## 6. Realism, Reference, and Rationality

Hilary Putnam (1982) has presented several arguments against the possibility of naturalistic or evolutionary epistemologies. One argument is that evolutionary epistemology presupposes metaphysical realism which, he claims to have shown, is incoherent. A second, more general argument is that naturalistic epistemologies attempt to eliminate normative reason. But reason, being both "immanent" and "transcendent," cannot be eliminated without committing "mental suicide" (1982, 22). Just explicating these arguments, let alone refuting them, would be a major undertaking. Here I can only attempt to locate some main points of disagreement and suggest where I think Putnam goes wrong.

Let us adopt Putnam's simple characterization of metaphysical realism as the view that "there is exactly one true and complete description of 'the way the world is'" (1981, 49). Must the evolutionary epistemologist or naturalistic philosopher of science make any such supposition? I do not see why. The naturalistic position is that our cognitive capacities are an evolutionary development of those possessed by lower primates and other animals. It is these same capacities the naturalistic philosopher of science employs in attempting to study the scientific activities of his fellow humans. Surely our primate ancestors could not be accused of being metaphysical realists. In so far as our cognitive abilities are continuous with theirs, why should we be any different? Perhaps some evolutionary epistemologists have indeed espoused metaphysical realism, maybe even claiming evolutionary support for such a position. But this is surely no necessary feature of an evolutionary perspective in epistemology.

In Putnam's terms, my naturalistic philosopher of science might be called an "internal realist." But naturalistic philosophers of science holding internal realism are, Putnam claims, no better off. He sees such a view as an attempt of *define* "rationality" in terms of the use of evolved capacities. The suggested formula is: Rational beliefs are those arrived at using evolved capacities for forming beliefs. But this formula is either obviously false or vacuous depending on whether we include all beliefs or only the rational ones. Thus, Putnam concludes, "The evolutionary epistemologist must either presuppose a 'realist' (i.e., metaphysical) notion of truth or see his formula collapse into vacuity" (1982, 5).

Here Putnam assumes that one of the tasks of a naturalistic epistemology would be to provide a *definition* of rationality. But one of the main points of an evolutionary perspective is that there is no sharp boundary between animals and humans, and thus between irrational and rational. From an evolutionary perspective, different organisms deal with aspects of their environments in more or less effective ways. Doing science is one of the ways we humans deal with aspects of our environment. Turning our attention to that process itself, we should expect to find that, in various respects, some people are more effective than others. And we would seek to explain why and how this comes about. Attempting to draw a fundamental distinction between rational and irrational activities is itself not an effective way to understand science, or any other human activity.

Of course I do not deny that providing a characterization of rationality is a well-entrenched feature of epistemology. By defining man as the rational animal, Aristotle bequeathed to philosophy the task of discovering the essence of rationality. We have given up essentialism in biology. It is about time we gave it up in epistemology, and for similar reasons.

As noted above, Putnam also has more general arguments purporting to show why reason (and by implication epistemology and the philosophy of science) cannot be naturalized. One line of argument is that reason requires language, which requires reference, which cannot be naturalized. Moreover, reason and language necessarily involve *values,* which also cannot be naturalized. I could not begin to untangle these arguments here. The most I can do is point out that, if Putnam is correct, then there are genuinely *emergent* properties, for example, the property of being rational.[9] Somewhere along the line from fishes to philosophers there emerged fundamentally irreducible properties that science alone cannot explain.

Arguments against emergentism have been given by many philosophers, including, several generations ago, Putnam himself (Putnam and Oppenheim 1958). I shall not attempt to review them here. I only marvel that anyone could think these arguments refuted by an analysis of the possible reference of 'cat' and 'cherries' (Putnam 1981, ch. 2).

From a naturalistic perspective, the urge to find some essential difference between animals and humans is itself something to be explained. The evolutionary process produced a species of creatures that has spent much of its history denying its evolutionary origins. Why do humans keep insisting on their special (if not outright superior) place in nature? Psychologists, sociologists, and historians of religion have, in various guises, attempted to answer this question. What strikes me is how self-serving the emergentist program can be: humans arguing that humans are a breed apart. One wonders if the rejection of a naturalistic approach to the philosophy of science (and philosophy generally) does not serve a much narrower self-interest. If the philosophy of science is naturalized, philosophers of science are on the same footing with historians, psychologists, sociologists, and others for whom the study of science is itself a scientific enterprise. The most philosophers of science could claim is to be the *theoreticians* of a developing science of science on the model of theoretical, as opposed to experimental, physics. Would that not be status enough?

## 7. Models and Theories

As is clear from the form of the circle argument, a crucial test for any naturalistic theory of science is its account of *theory choice*. Since it would be impossible adequately to develop and defend a naturalistic account of theory choice in a short space, I will only present enough to show that such an account is both possible and at least somewhat plausible. Before one can discuss theory choice, however, it is necessary to say something about the objects of choice, namely, theories.

Since Euclid there has existed a more or less continuous tradition of representing theoretical knowledge in the form of an axiomatic system. Newton was part of this tradition, and so were the founders of modern logic. For most of this century, philosophers who have drawn their inspiration from logic and the foundations of mathematics have assumed that a theory is some type of formal, axiomatic system. The fact that scientists in the twentieth century rarely present theories in axiomatic form has not been very troubling because the philosopher's task has been seen as one of reconstruction, conceptual analysis, or justification—not description. If, however, one takes the descriptive task as fundamental, the axiomatic account clearly is not adequate. For the most part it is simply not true that theoretical scientists are engaged in developing axiomatic systems. This point is obvious for the major recent theoretical developments in sciences such as biology and geology, but it holds even for physics. Where are we to find a better account of scientific theories?

If we restrict ourselves to recent science (since 1900 or 1945), the task is easier because the transmission of theoretical knowledge has become quite uniform. It relies heavily on the advanced *textbook*. Until beginning dissertation research, most scientists in most fields learn what theory they know from textbooks (in conjunction with lectures, which also follow a textbook format). Thus, if we wish to learn what a theory is from the standpoint of scientists who use that theory, a good way to proceed is by examining the textbooks from which they learned much of what they know about that theory.

Classical mechanics provides a good example. Many sciences were modeled on mechanics and borrowed heavily from its mathematical techniques. And for many scientists and engineers today, classical mechanics provides their first experience with a real theory. In addition, classical mechanics has been a standard example for philosophers advocating an axiomatic account of theories. It thus allows a direct comparison of the merits of any rival account.

Looking at typical upper-division or graduate-level texts, what do we find? Often there is a chapter of mathematical preliminaries. The first substantive chapter, however, almost invariably presents Newton's three laws of motion. To apply these laws, one needs a force function. The following chapters, therefore, are typically devoted to the use of Newton's laws of motion with various force functions. A not too advanced text might devote a chapter to uniform forces—Galileo's problem of falling bodies. A typical next chapter takes up the case of a linear restoring force in one dimension, Hooke's Law—which yields a linear harmonic oscillator. Later one meets the Law of Universal Gravitation, the inverse-square force that yields orbital motion in two dimensions. And so on.

Within each chapter one finds, among other things, the following: (i) mathematical solutions to the equations of motion incorporating the specific force function at issue; and (ii) examples of kinds of real systems to which these particular equations of motion might be applied. One learns, for example, that a linear restoring force yields a sinusoidal motion, and that the horizontal motion of a pendulum is approximately sinusoidal.

One of the most significant other things one learns is that none of the systems cited as examples *exactly* fit the equations. The horizontal restoring force of the pendulum, for example, is only linear in the limit as the angle of swing approaches zero. Regarding the equations as straightforward statements which are either true or false is, therefore, bound to misrepresent the situation. How, then, should we represent it?

I suggest we take the equations as characterizing an abstract, idealized system, for example, the simple harmonic oscillator. Calling such a system a "model" (or theoretical model) agrees pretty well with both scientific and philosophical usage. Claims about real systems, then, have the form: the real system *is similar to* the model. A pendulum with small amplitude, for example, is similar to a simple harmonic oscillator. I will call such claims "theoretical hypotheses." Implicit in any theoretical hypothesis is a specification of the respects and degrees in and to which the similarity is claimed to hold. At this point one could introduce truth and falsity for theoretical hypotheses, but a claim of truth here would be redundant, serving only to facilitate semantic assent.

The typical advanced text, then, presents the student with a *cluster of models* (really a cluster of clusters) together with a number of hypotheses about real things claimed to be similar to one or another of the models. For the purpose of developing a naturalistic theory of science, I suggest we understand the word "theory" as including both the cluster of models and a broad range of hypotheses utilizing these models. Restricting "theory" either to the models or to hypotheses produces too great a variance with how scientists use the term. For all sorts of reasons, it is best to stick as closely to scientific usage as is compatible with developing an overall, adequate theory of science.

In working through a standard text, students learn many things that are best not regarded as explicitly part of the theory, but that are very important nonetheless. They learn the accepted *interpretation* of general terms such as "position," "mass," and "force." They also learn how to *identify* particular positions, masses, and forces. Any theory of science must assume that scientists have the ability to make these sorts of interpretations and identifications. Securing a better understanding of how this is done, however, can safely be left to linguistics or, more generally, to the cognitive sciences.

It is evident that the above account of theories is realistic without going to the extreme of "metaphysical realism." Indeed, it is compatible with some recent forms of anti-realism. I would call it "constructive realism." It is "constructive" because models are humanly constructed abstract entities. It is realistic because it understands hypotheses as asserting a genuine similarity of structure between models and real systems without imposing any distinction between "theoretical" and "observational" aspects of reality. It is not "metaphysical" in that it makes no claim that there is one true and complete description of any real system. A constructive realist need not claim, for example, that there is a uniquely correct classical model for describing any actual pendulum. Nor must one claim similarity with the real world for *every* aspect of a model. One can be selective in choosing those respects in which the similarity is claimed to hold.[10]

From a naturalistic perspective, then, the theory of classical mechanics appears not to have the structure of an axiomatic system. At best an axiomatic structure could be imposed on one particular type of model, for example, systems of particles subject only to inverse-square forces. Nor, contrary to Popper's philosophy, do *universal* statements play a major role. No longer does one find sweeping Laplacian generalizations about all bodies in the universe. The typical hypothesis only asserts a similarity between a model and a more or less restricted class of real systems such as pendulums. There are many more lessons to be learned from a serious study of science textbooks, but these are sufficient to proceed to a sketch of naturalistic theory choice.

## 8. Naturalistic Theory Choice

In the philosophical literature, the problem of theory choice has almost universally been understood as one of characterizing *rational* choice. Most philosophers have been willing to grant that it would be rational to choose theories that are true (or at least approximately true). The trouble is, of course, that we do not have an independent check on which are the true ones.

Philosophical treatments of theory choice, therefore, have generally proceeded by focusing on properties other than truth, and then attempted to establish a general principle saying it is rational to choose theories with the specified properties. Among the many properties of theories suggested for this role have been: simplicity, falsifiability, high degree of logical probability, high degree of corroboration, predictive power, explanatory power, and fruitfulness. The preferred way of establishing the required general principle is by demonstrating a connection between the specified properties and truth. Despairing of establishing any such connection with truth, however, many philosophers have argued for the rationality of theory choice in terms of these other properties themselves.

The post-Positivist switch to larger units of analysis (paradigms, research programs, or research traditions) has not significantly changed the general strategy. The difference is that now one focuses on properties of the larger unit, such as progressiveness, and then argues that it is rational to choose a tradition with these properties. The choice of individual theories is subordinated to the choice of the corresponding larger unit.

All of these approaches assume the more general principle of rationality that scientists generally strive to make a *rational* choice, however this is defined. Other than this general principle, philosophical accounts of theory choice make scant reference to the actual flesh-and-blood scientists who do the choosing. The approach is almost totally "top down." A naturalistic approach to theory choice is explicitly "bottom up." It begins with real agents facing various choices in the course of their actual scientific lives. It assumes that choosing theories is not too dissimilar from choosing anything else, and then looks at how humans in fact make choices.

If our naturalistic theory of science is not to be *merely* historical, we need a *theory* of theory choice. I would suggest that decision theory includes some models of choice that can provide at least a start. Decision theory, however, has a split personality. Sometimes it operates as an account of *rational* choice; other times it is more descriptive. Here we want the descriptive mode, which may be viewed as a specialized part of ordinary belief-desire psychology.

Taken descriptively, decision-theoretic models begin with a *decision problem* that may be represented as a matrix defined by a set of possible options and a set of possible states of the world. The agent's *desires* (or values) are represented by a ranking or utility measure over the option-state pairs, the *outcomes* of the decision process. The result of adding the agent's values is a completed value (or "payoff") matrix. The role of the agent's beliefs in decision making is more complicated, as will be illustrated below.

The focus of rational decision theory has always been on the *decision rule* (or decision strategy) that defines *the* rational choice as a function of the payoff matrix. The problem of rational decision theory has been to establish a uniquely rational decision strategy. Descriptive decision theory looks instead at the characteristics of the decision strategies that are actually used.

Among the most promising descriptive strategies is *satisficing*. Agents following a satisficing strategy must have a good idea of their minimum satisfactory payoff—their satisfaction level. They then survey their options to see whether any have at least a satisfactory payoff for each possible state of the world. If such an option exists, that is the one chosen. If there is no satisfactory option, agents must either lower their satisfaction level or otherwise change the decision problem. Following a satisficing strategy thus guarantees at least a satisfactory payoff—unless, perhaps, no decision is made.[11]

One could, of course, go on to argue that satisficing is rational. But there is no need to do so. Rather, we can take the satisficing strategy as part of our theoretical model of human decision making.

We can then investigate the characteristics of the model and inquire of the circumstances, if any, in which humans fit this model. The fit need not be perfect. Like many theoretical models, this one is highly idealized. My hypothesis is that scientists typically follow something approximating a satisficing strategy when faced with the problem of choosing among scientific theories. If this is correct, we have a good scientific explanation of theory choice in science.

## 9. A Role for History

Kuhn was of course correct in thinking that a naturalized philosophy of science would provide a role for history. The role, he suggested, was as *evidence* for theoretical claims about science. Yet the use of history by philosophers of science (recall the metamethodology of Lakatos and Laudan) suggests that this evidential relationship is more complex than it might seem.

It is useful to consider how some other sciences use the historical record as evidence for their theories. Evolutionary biology and economics provide appropriate models because they seem nicely to bracket a proposed theoretical science of science. Turning first to evolutionary biology, it is generally thought that the fossil record provides historical evidence for evolutionary theory. I am far from convinced that this record, by itself, provides a satisfactory basis for deciding that any evolutionary theory is correct. Here, however, I am concerned with a narrower issue. Those who have used the fossil record as evidence for evolutionary theory have generally assumed that the underlying mechanisms of evolution, whatever they might be, are relatively stable. Few biologists have ever argued that we might need different models of evolution for different epochs. The major recent controversies over punctuated equilibria or mass extinctions concern the nature and rate of changes in the environment—not our models of the underlying evolutionary mechanisms.

In contrast to evolutionary biology, the most successful theoretical models in economics, whether macro or micro, are *equilibrium* models. The data for such models are, therefore, not historical in the sense that they follow economic developments over time. For models that do use genuinely historical data, one must turn to theories of economic *development*. These, however, are generally thought to provide the least successful models in all of economics, Marxism being the most obvious example. The generally accepted reason for the poor record of models of development is that the economic mechanisms themselves change over time. It is not simply a matter of looking at the same mechanisms operating in a different environment. A rural, agrarian society, for example, seems to embody different economic mechanisms than an urban, industrial society.

Following Kuhn, historically minded philosophers of science have argued, using historical examples, that not only the content of science changes with time. Its aims and methods change as well. This seems to imply that the relation between theories of science and the history of science follows the economic rather than the biological pattern. Indeed, the Kuhnian model of development—normal science, crisis, revolution, new normal science—seems to have as much, or as little, theoretical content as the Marxian stages of economic development. One wonders whether any theory of scientific development that includes changes in aims and methods could do much better. Yet most historically oriented philosophers of science since Kuhn seem to be aiming at a similarly grand theory of development. Few would describe their own aims in these terms, of course. But illustrating the same point using historical cases ranging from Newton, through Lavoisier, to Einstein, and even J. D. Watson, betrays the intent.

The options for a naturalistic theory of science, then, are these. The first is an ambitious strategy that seeks mechanisms of scientific development that can explain not only changes in content but also changes in aims and methods. One could then claim to have similar mechanisms operating over long periods, say from the seventeenth century to the present. The danger in this strategy is ending up with only vaguely defined models of science. A second, much less ambitious strategy would be to restrict attention to shorter epochs such as science in the seventeenth century or since World War II. Here the danger is ending up with models of only very restricted applicability.[12]

Following my own theory of science, I would suggest a third, hybrid strategy. Begin with the less ambitious strategy, and then try to link up the various models so as to obtain a cluster of partially overlapping models covering several epochs, perhaps, even, most of science since Newton. That, I suspect, is the most that can be done. The suggested model of science provides some hope for thinking that the third strategy can be successful. The activities of model construction and model choice abstract from the scientific context in much the same way as models of mutation and selection abstract from the biological environment. These activities may take place in many different social and economic settings. Different aims, or values, may be reflected in the structure of decisions concerning specific hypotheses. So may the information yielded by new methodologies. Whether this is enough to provide informative similarities among widely separated epochs remains to be seen.

My aim in this paper, however, has not been to argue for a particular strategy or a particular model of science. These have been noted only to illustrate the possibilities opened up by a naturalistic approach. The main thesis is that the study of science must itself be a science. The only viable philosophy of science is a naturalized philosophy of science.

# NINE

# Constructive Realism

## 1. Introduction

Empiricism, writes van Fraassen, "could not live in the linguistic form the positivists gave it." His own empiricist image of science, therefore, utilizes an alternative linguistic form for scientific theories. In liberating empiricism from its positivist shackles, however, van Fraassen has unintentionally also set free the realism he abhors. That, at least, is the thesis of this chapter.[1]

*The Scientific Image* treats a number of topics that would be central in any theory of science. I shall focus on just two: (1) the nature of scientific theories and their relations with the world, and (2) the justification, or acceptance, of theories. Van Fraassen devotes much more attention to the nature of theories than to their justification. It is his account of what theories are that liberates both empiricism and realism. His arguments for empiricism over realism, however, turn primarily on questions of justification. By adopting the core of his view about theories, I am in the enviable position of being able to employ his own best-developed weapons to attack his weakest defenses.

I would not enter this battle if I did not feel strongly that realism is right and empiricism wrong—and, moreover, that the difference matters. It matters because I take our task to be not merely to engage in scholastic debate, but to construct a general *theory of science* that could provide a theoretical background for diverse studies in the history, philosophy, psychology, and sociology of science.

## 2. Models

According to van Fraassen, the logical empiricists' preoccupation with the linguistic structure of scientific theories obscured the importance of the *models* satisfying those linguistic structures. In science, he claims, it is the models, not the linguistic forms, that occupy center stage. I agree completely.

Van Fraassen draws much of his inspiration from quantum theory. His image of theories, however, is intended to be much more general, and he does employ classical examples. We should not go too far astray, therefore, if we stick to classical mechanics. Consider, then, a one-dimensional linear harmonic oscillator—one of the original exemplars of Newtonian science.

As described in texts for nearly three hundred years, a linear harmonic oscillator is a mass subject to a restoring force proportional to the distance from its rest position. For reasons that will be obvious shortly, I agree with van Fraassen in preferring the Hamiltonian formulation:

> A one-dimensional linear harmonic oscillator is a system consisting of a single mass constrained to move in one dimension only. Taking its rest position as origin, the total energy of the system is,
>
> $H = T + V = p^2/2m + 1/2\, kx^2$ where $p = m\, dx/dt$.
>
> The development of the system in time is given by solutions to the following equations of motion:
>
> $dx/dt = \partial H/\partial p \qquad dp/dt = -\partial H/\partial x.$

This is a capsule version of descriptions found in standard textbooks (see, for example, Marion 1970, ch. 3).

Thinking of **k** and **m** as merely constants, and both **x** and **p** as simply mathematical functions of **t**, solutions of the equations in **x-p-t** space are elliptically shaped spirals moving out along the **t** axis.

Projecting the full **x-p-t** state space onto the **x-p** plane yields an ellipse. Projections onto the **x-t** and **p-t** planes yield sinusoidal curves. Abstracting, then, from the standard meanings of terms like "mass" and "position," we are left with a family of purely *mathematical* structures determined by the parameters **m** and **k** (and by specified values of both **x** and **p** for some given value of **t**).

Van Fraassen follows Patrick Suppes in holding that the proper language for the philosophical study of science is mathematics, not metamathematics. This doctrine frees the philosophy of science from the concerns and methods of the foundations of mathematics, one of the twin original sources of inspiration for Logical Empiricism—the other being the classical empiricism of Hume, Mill, Russell, and Mach. Suppes, however, retained the idea of a canonical language, namely, set theory. Van Fraassen frees us from even this constraint, letting the appropriate language be dictated by the specific scientific subject under investigation. On his view, interesting foundational problems in the various sciences are generally not such that they can be removed merely by reformulation in the proper linguistic framework. They reside in the structure of the models employed.[2] Van Fraassen has another reason for rejecting the set-theoretic framework. It is strongly biased toward purely *extensional* formulations of scientific theories, and van Fraassen wants to represent the *modalities* as well. I will return to this point later.

The desire to free philosophy of science from general questions about language is laudable. Philosophy of science should no more be just a branch of the philosophy of language than of epistemology or the foundations of mathematics. One cannot, however, eliminate all questions about language or *interpretation*. Any consistent formal structure has purely mathematical models, say in number theory. Some additional semantic categories, such as meaning or reference, are needed to distinguish masses from numbers—and thus mechanics from pure mathematics.

In earlier publications, van Fraassen introduced "elementary statements" which yield what he then called a "semi-interpreted language" (1970b). Interpretation of purely mathematical symbols is thus provided, or at least transmitted, by the elementary statements. I will simply ignore such issues here. The theory of science need not wait on the development of adequate general theories of meaning and reference to proceed. We need not know in detail *how* general terms such as "mass" come to be associated with terms in an abstract mathematical structure. We know that it *can* be done because it *is* done.

Assuming standard interpretations of the basic terms, then, we obtain a generalized model (or family of specific models) that we call *the* linear harmonic oscillator. I will use the term *theoretical model* to refer either to a general model or to one of its specific versions obtained by specifying unique values for all parameters and initial conditions. A theoretical model, then, is a *defined* entity. It has all and only those characteristics explicitly specified. In cases where the definition employs mathematical concepts, the model has all the precision of the corresponding mathematical notions. The mass in *the* linear harmonic oscillator, for example, exhibits a perfectly sinusoidal motion. There can, therefore, be no question whether simple harmonic oscillators perfectly satisfy the equations of motion. That they do is a matter of explicit definition. On the other hand, if the behavior of the theoretical model is a matter of definition, what is the relationship between theoretical models, so conceived, and real oscillating systems such as bouncing springs, pendulums, and vibrating molecules?

## 3. Theoretical Hypotheses

The logical empiricist notion of a *correspondence rule* conflated what on a model-based view are two distinct, though related, functions. One is providing a general *interpretation* of theoretical terms such as "mass," "momentum," etc. The other is providing the means for *identifying* particular instantiations of these terms. I think van Fraassen would agree that the details of this process are not a matter for philosophical analysis or for armchair psycholinguistics. Rather, the topic calls for deep empirical investigations into how we humans use abstract symbols in describing particular objects in the real world. Here again we do not need a detailed account at hand to construct a useful theory of science. A fairly shallow analysis will do.

Figure 9.1 pictures a weight suspended on a spring—the kind of

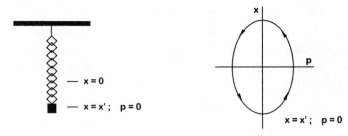

9.1   A bouncing spring and the state space for an associated model.

thing one finds in elementary physics laboratories. Also pictured is the two-dimensional **x-p** state space with initial conditions specified. The arrows indicate the evolution of the state of the system in the state space. Here we make the "standard" identifications. The object on the spring is our mass. The place the mass comes naturally to rest is taken as the origin of our coordinate system. Moving up vertically is the positive **x** direction. We shall assume that the system is set in motion by pulling the weight down a distance **x'** and releasing it with no initial velocity.

In the circumstance indicated, what is the relationship between our *theoretical model,* the ideal linear harmonic oscillator, and this real oscillating system? Here, finally, we confront the difference between constructive empiricism and constructive realism. Let us begin with realism. At one point, van Fraassen characterizes the realist as claiming "to have a model which is a faithful replica, in all detail, of our world" (1980, 68–69). In all fairness, this is not his most considered statement of a realistic hypothesis, but it provides a useful starting point.

Now, our theoretical model is obviously not a "faithful replica, in all detail," of our bouncing weight. Real springs, for example, have internal friction which eventually brings the system to rest. And there are other apparent forces, centrifugal and coriolis, due to our reference system being on a rotating sphere, i.e., the Earth. These circumstances insure that the motion of the actual system will differ from that indicated by the model.

The traditional response to such worries is to construct a more detailed model. So the claim would be that there is *some* Newtonian model that *exactly* captures the behavior of the real system. Since Einstein, however, we know that this claim is false. There is *no* Newtonian model that exactly represents any dynamical system. Moreover, even in the classical context, say as of 1875, such a claim could not have been *justified*. Since all evidence is finite and of less than perfect precision, the most that could ever have been justifiably claimed is that there is some Newtonian model that matches the real system to within then-existing experimental limits of detection. Here I am invoking the principle that one cannot justify a claim by appeal to experiments that have no chance of detecting the falsity of the claim even if it is false. Later I will provide a basis for this principle.

If we are to have scientific hypotheses which are realistic and also have some reasonable chance of being true, we must avoid claims that any real system is exactly captured by some model. Realists typically have adopted this view. Most realists would qualify their claims by saying that the real system is at best *approximately* captured by the model. The difficulty with this formulation is to explain what "approximately" means.[3] Van Fraassen suggests that approximation be explained in terms of a *class* of models, one of which *exactly* fits (1980, 9). This explication, however, must be rejected. Since we now believe that there is no Newtonian model that exactly captures any real system, van Fraassen's analysis would have us deny that Newtonian models now provide a good approximation for any real world systems. Such a position is just too much at variance with the present convictions of scientists—and of most students of scientific practice.

I propose we take theoretical hypotheses to have the following general form:

> The designated real system *is similar to* the proposed model in specified *respects* and to specified *degrees*.

We might claim, for example, that all quantities in our spring and weight system remained within ten percent of the ideal values for the first minute of operation. The restriction to specified respects and degrees insures that our claims of similarity are not vacuous.

The precision associated with any hypothesis cannot justifiably be *greater* than that of the measurement techniques actually employed, though one might in many contexts claim *less* precision than could be achieved. I shall say little more about precision here except to note that this is one example of how the logical empiricists' preoccupation with simple languages obscured what almost everyone knows to be an essential feature of scientific hypotheses. The state-space representation makes possible an intuitive analysis of degrees of approximation in terms of a volume in the state space which includes the idealized history of the theoretical model.

The above general characterization of a theoretical hypothesis permits a number of variations corresponding to various grades of empiricism and realism. The most extreme realist hypothesis would be that the real system resembles the model in *all* respects. There are many examples showing that this version of realism is too extreme. Here one might mention the phase factor in the wave functions of quantum theory, the imaginary part of complex numbers used in electromagnetic theory, or even negative square roots.

I would therefore recommend a more modest, constructive realism that claims only a similarity for *many* (or perhaps most) aspects of the model. This formulation leaves the residual problem of saying when an aspect of the model is to be denied a counterpart in reality. The discovery of the positron shows us that it sometimes pays to take seriously a negative square root—and thus that there is unlikely to be a simple answer to the residual problem.

How, then, does the above modest realism differ from van Fraassen's constructive empiricism? At first sight, the difference seems obvious. Empiricism is much more restrictive. It enjoins us to limit our claims of approximate similarity to *observable* aspects of our models. The difference between realism and empiricism, therefore, turns on how one specifies the observable aspects.

Van Fraassen is clear in denying the logical empiricist doctrine that the specification of observable aspects can be done in terms of vocabulary or by segregating individual statements. Rather, it is done by designating substructures of one's models as the *empirical substructures*. Empiricists claim only that a model is *empirically adequate*—meaning that there is some empirical substructure which approximately matches the observable facts. How, then, do we pick out the empirical substructures? Here van Fraassen is uncharacteristically unclear in that his examples and theoretical pronouncements do not always cohere.

He considers, for example, Newton's claims about absolute space. Imagine, then, that we complicate our model for the simple harmonic oscillator by adding a constant velocity representing the motion of the solar system relative to absolute space. No matter what velocity we add, the resulting behavior of the system, as measured in our Earth-based coordinate system, is unchanged. It is not just that we can't observe any difference. Our model itself tells us that there is no difference we could detect, no matter how sophisticated or how sensitive our instruments. Here the constructive realist could agree completely with van Fraassen that to assert any similarity between the real system and the model with respect to some added absolute velocity term would be gratuitous and totally unjustified.

Van Fraassen, however, appears to claim more. Indeed, he seems to endorse the idea that the appropriate empirical substructures for classical mechanics consist only of relative positions and velocities. Even the standard Hamiltonian state space, which includes momentum, and thus mass, is too theoretical. Why? I am not sure. The kinds of reasons that have been given by other philosophical students of classical mechanics make strong appeal to the logical structure of mechanics itself.[4] Van Fraassen, however, explicitly denies that his distinction is theory-dependent in this way.[5]

The correct approach, he suggests, is not through philosophical analysis or armchair psychology, but through the empirical study of human perceptual capabilities. So determining what is observable depends on scientific theory, but on psychology and physiology, not on physics. Whether this generates a "logical catastrophe" for psychology and physiology is unclear.[6] But it is very difficult to see how the study of human perceptual capabilities could tell us that velocity is observable while mass is not.

The operative scientific notion, I suggest, is not human observability but scientific *detectability*. What is now detectable depends primarily on two things: (1) the structure of our current models, together with our interpretations and identifications, and (2) our present ability to design and build experimental apparatus, which in turn depends on other models. That the ultimate output of our measuring instruments must be something that *humans* can *observe*, e.g., a dial or a computer printout, is a simple consequence of the fact that all the scientists we know about are humans. Satisfying this requirement is primarily a matter of engineering.[7]

If the requirement of empiricism is only that our scientific claims be restricted to aspects of our models that are, in a broad sense, *detectable,* then one major difference between constructive empiricism and constructive realism is removed. There remain serious differences regarding *modalities,* that is, what our models tell us about the *possible* behavior of real systems. This is the subject of the following section. To sum up the present section, van Fraassen is to be thanked for liberating us from one of the two sources of Logical Empiricism, metamathematics and formal linguistic analysis. But he remains in the clutches of the other source—classical empiricism. His empiricism, therefore, is a vestige of the classical empiricist philosophy that sought to ground all knowledge in *experience*—and thus to make humans the *measure,* rather than merely the *measurers,* of all things. Such an empiricism may indeed "deliver us from metaphysics" (van Fraassen 1980, 69), but it delivers us into the hands of an anthropocentrism that is antithetical to the whole modern scientific tradition.

## 4. Physical Modality

Most recent versions of the debate between empiricists and realists have focused on the distinction between observational and theoretical aspects of science. Van Fraassen's treatment is no exception. He does, however, give special consideration to the status of the physical *modalities.* This seems to me the crucial dividing line between empiricism and realism. Van Fraassen is willing to grant that real systems may in fact exhibit the *theoretical* structure of our models. He merely insists that we cannot justifiably assert this correspondence. Regarding physical *modality,* however, he is not merely agnostic, but atheistic. Possibilities and necessities are only figments of our models—useful, perhaps, but not even candidates for reality.[8] Thus, even if we substitute detectability for observability, there remains a vast difference between empiricism and a realism that extends to physical modality. Here I will merely describe the position of modal realism. The following section will treat the justifiability of modal claims.

Van Fraassen's own preferred form for presenting theories, the state-space approach of Hermann Weyl and Evert Beth, provides an excellent framework for expressing the claims of modal realism. To keep the discussion as concrete as possible, let us return to the harmonic oscillator and bouncing weight represented in figure 9.1. I will begin by distinguishing a number of possible claims, of increasing strength, about the relationship between the idealized model and the real system. In each case, an appropriate degree of precision over a suitably restricted interval of time is assumed.

1. (Extreme empiricism) The model agrees with the positions and velocities of the real mass which have been observed up to the present time.
2. (Extended empiricism) The model agrees with all the positions and velocities that ever have been or will be observed.
3. (Actual empiricism) The model agrees with all the actual positions and velocities of the real system, whether they are observed or not.
4. (Modal empiricism) The model agrees with all possible positions and velocities of the real system.
5. (Actual realism) The model agrees with the actual history of all (or most) system variables.
6. (Modal realism) The model agrees with all possible histories of all (or most) system variables.

Constructive empiricism is a version of actual empiricism. One could identify constructive realism with either actual realism or modal realism. I prefer the modal version. In either case, constructive realism is a variety of what is sometimes called *structural realism*.[9]

As Quine delights in pointing out, it is often difficult to *individuate* possibilities. Often, yes, but not always. Many models in which the system laws are expressed as differential equations provide an unambiguous criterion for individuating the possible histories of the model. They are the trajectories in state space corresponding to all possible *initial conditions*. Threatened ambiguities in the set of possible initial conditions can be eliminated by explicitly restricting the set in the definition of the theoretical model. If, for example, one wonders whether $x' = 100$ meters is a possible initial condition, we could explicitly restrict $x$ to $\pm 10$ cm. Nor is this criterion limited to models defined in terms of differential equations. A clear distinction between system laws and parameters or initial conditions is generally sufficient. Of course, even classical physics presents cases in which the specification of boundary conditions is not unambiguous. But the ambiguity is not nearly so great as Quine or van Fraassen suggest.

As noted earlier, part of van Fraassen's reason for preferring the state-space representation of theoretical models, rather than a purely set theoretic representation, is that the state-space representation emphasizes the *modal* structure. Thus, even for van Fraassen, the modal structure is important for our understanding of reality. And, indeed, for a two- or three-dimensional state space, one can easily visualize the possible histories. But van Fraassen explicitly denies that the modal structure is, or even could be, a part of the physical world.

One's attitude toward modalities has a profound effect on one's whole theory of science. Actualists, including actual realists, must hold that the aim of science is primarily to describe the actual history of the world. For modalists, including modal empiricists, the aim is to describe the structure of physical possibility (or propensity) and necessity. The actual history is just that one possibility that happened to be realized. This difference in aims is connected with profound differences in how one understands diverse scientific activities such as causal attribution, explanation, and experimental design.

Our bouncing spring is a *causal* system if anything is. Now, our theoretical model tells us that the frequency of oscillation, $f$, is *functionally* related to the ratio, $k/m$, and functionally independent of the amplitude, $x'$. If the functional relationship obtains merely between the *actual* values of these quantities, it is difficult to see what more there is to *causality* than merely this functional relationship. Empiricists have traditionally sought to ground the causal claim in *universal generalizations,* e.g., all bouncing springs exhibit these relationships. On examination, such generalizations turn out to be either false or vacuous. For the modal realist, the *causal* structure of the model, and thus, to some degree of approximation, of the real system, is identical with the *modal* structure. For any real system, the functional relationships among the actual values of $p$, $m$, $k$, etc., are causal not because they hold among the *actual* values in *all* such real systems, but because they hold among all the *possible* values in *this* particular system.

Giving *explanations* for particular happenings or citing causes thereof is, as van Fraassen insists, a complex human activity involving theoretical hypotheses. It is thus not to be expected that the correct responses to requests for explanation or for causal antecedents are uniquely determined by the scientific hypotheses themselves. It does not follow, however, that such questions do not have unique answers when applied to *restricted* systems such as our weight and spring, for which we have a complete model. In such cases, there is no need for externally imposed relevance conditions. The model itself tells us everything that could be relevant to a system exemplifying its structure. And it gives unique answers for any fully specified set of parameters and initial conditions. What would the system do if we were to change the initial conditions in a specified way? Our hypothesis gives us a unique answer. The uniqueness is lost only when the physical system is embedded in a larger social context for which we have no similarly complete models.

From the standpoint of modal realism, to *understand* a system is to know how it works. And this means knowing how it would behave under conditions other than those which in fact obtain. It is knowing the causal structure. In a curious way, this is also van Fraassen's view. The curiosity is that he would replace "knowing" with "having an empirically adequate model with the given modal structure." But this reduces scientific understanding derived from a theoretical model to understanding conferred by a good historical novel, one which remains faithful to the known historical facts. They both provide a good story, which, however, we have no reason to believe is true. I am not sure what motivates van Fraassen to advocate such a degree of epistemic caution, but I am fairly confident that it is not justified.

## 5. Acceptance

Van Fraassen distinguishes between the empiricist *acceptance* of a model as empirically adequate and the realist *belief* that the model provides a true picture of reality. I shall speak of the acceptance of hypotheses in both cases. The difference is in the nature of the hypotheses at issue. Empiricist hypotheses claim only that a real system resembles the model in its observable aspects, while realist hypotheses assert the resemblance for theoretical (and perhaps modal) aspects as well.

In *The Scientific Image,* the main argument for empiricism is that realism is unnecessary. Everything we want to say about science, e.g., its use in giving explanations, can be said on an empiricist understanding of scientific hypotheses. I shall take up this argument, though in an indirect way, in the conclusion to this chapter.

Here I wish to confront a more direct, secondary argument. This argument is based on the logical principle that if **H** implies **H'**, then, no matter what the evidence, **H** cannot be better justified than **H'**. In other words, evidential *safety* is inversely related to logical *strength*. Thus, even if one has other reasons for wanting to be a realist, one's realistic hypotheses could never be better justified than the corresponding empiricist hypotheses.[10] The obvious reply is that one might wish to trade off some evidential security for the logical strength of realistic hypotheses. And van Fraassen cannot totally deny this appeal, since he himself stops short of embracing extreme, or even extended, empiricism. But the issue deserves a deeper investigation.

I shall argue not only that realistic hypotheses can be justifiably accepted but also that, in some circumstances, a realistic hypothesis might be *better* justified than the corresponding empiricist hypothesis. The argument requires appeal to an explicit theory of acceptance. Such an appeal carries no danger of directly begging any questions, since van Fraassen does not himself present an account of acceptance.

Given his generally pragmatic approach to epistemic issues, van Fraassen would agree that the activities of scientists are to be understood in essentially the same way as the activities of other people engaged in other pursuits. Now, our everyday understanding of human behavior is in terms of beliefs and desires. That is, we generally explain why people do what they do in terms of their desires, or goals, and their beliefs both about these goals and about various possible means to their goals. Moreover, there exist fairly well developed theoretical models for describing such behavior. These are supplied

|  | H is true | H is false |
|---|---|---|
| Accept H | V = 1 | V = 0 |
| Reject H | V = 0 | V = 1 |

9.2 A decision-theoretic representation of the acceptance or rejection of a given hypothesis.

by decision theory. Like all theoretical models, decision-theoretic models are only partial and approximate. And there are other kinds of models one might apply.[11] My aim here, however, is only to show that my approach to the justification of realistic hypotheses is sufficient, not that it is the best possible.

Although he does not present a theory of acceptance, van Fraassen seems to hold an implicit view of what such a theory must do. Recast in decision-theoretic terms, van Fraassen's view is that we must choose one hypothesis out of a literally countless set of possible hypotheses. This looks to be a very difficult task given finite evidence and the logical underdetermination of our models by even their complete empirical substructures. I would urge a quite different picture.

Looking at the scientific process, we see scientists struggling to come up with even one theoretical model that might account for phenomena that previous research has brought to light. In some cases we may find two or three different types of models developed by rival groups. These few models may be imagined to be a selection from a vast number of logical possibilities, but the actual choices faced by scientists involve only a very few candidates. Given this already limited context, we sacrifice little by assuming that the choice is simply between a hypothesis and its negation.

The simplest decision-theoretic representation of the acceptance or rejection of a given hypothesis is a two-by-two matrix, as pictured in figure 9.2. The two possible actions are "Accept **H**" and "Reject **H**." The two relevant possible states of the world are that **H** is true and that **H** is false. These descriptions of the actions and states require some immediate qualifications. To accept **H** means to regard it as being true. But acceptance is not absolute. It is provisional in the sense that it may be revoked in the light of new evidence. And even without further evidence, one acknowledges the *possibility,* indeed, some *probability,* that **H** is really false. We are all fallibilists.

**H** has the form earlier proposed for theoretical hypotheses, namely, that the real system is similar to the model in specified respects and to specified degrees. We can therefore use the same matrix for both empiricist and realist hypotheses simply by changing our understanding of what sorts of respects are included. The difference between the empiricist and the realist, in this framework, shows up as a difference in *goals.* The empiricist strives to accept true empiricist hypotheses; the realist strives to accept true realistic hypotheses.

Any decision-theoretic model requires a value input. A full treatment of the decision problem could thus be quite complex, since there is in reality a variety of relevant values associated with the individual, the particular field, and even the institution of science as a whole. For our simplified model of the situation, we can get by assuming only that truth is valued over falsity. We need not even attempt to distinguish the relative value of the two kinds of possible mistakes. Giving correct acceptance or rejection the value **1** and incorrect acceptance or rejection the value **0,** we obtain the completed matrix shown in figure 9.2.

A decision-theoretic framework also requires the adoption of some explicit *decision rules.* I agree with van Fraassen in shying away from Bayesian models of scientific agents. Such models assume scientists to have a belief function over hypotheses, a function having the formal properties of probability. There is little evidence that scientists actually have such nicely ordered beliefs. For this and other reasons, I prefer a version of classical decision theory. In all honesty, I think van Fraassen could better defend his position from a Bayesian standpoint. It is a tribute to his sensitivity to the practice of science that he does not take this route. Virtue, however, is not always rewarded.

From the standpoint of classical decision theory, the decision problem represented in figure 9.2 has no solution. The matrix is completely symmetrical. The symmetry, of course, is partly the result of our having assigned equal value to the two types of successful and mistaken decisions. But this is not essential. No one would argue that

|        | H is true | H is false |
|--------|-----------|------------|
| (A, R) | V = 1     | ?          |
| (R, A) | V = 0     | ?          |
| (A, A) | V = 1     | V = 0      |
| (R, R) | V = 0     | V = 1      |

9.3  A matrix for the metadecision problem.

**H** should be accepted or rejected solely on the basis of such value differences. The reason the problem is indeterminate is that it is *incomplete*. We have not yet factored in any *evidence*. Both empiricists and realists agree that the decision is to be made in light of the evidence. The question, therefore, is how the role of the evidence is to be represented in our model of the decision problem.

Evidence enters the decision problem by way of the *decision rule*. Although there are in principle any number of possible rules, the following simple rule is consistent with van Fraassen's presentation: (1) Accept the hypothesis if the known observations are as implied by the hypothesis; reject the hypothesis if the known observations are inconsistent with it. To evaluate this rule, we must compare it to other possible rules. There are only three other rules with the same general form. These are: (2) Accept the hypothesis if the observations are inconsistent with it; reject the hypothesis if they are as it implies. (3) Accept the hypothesis whether or not the observations are consistent with it. (4) Reject the hypothesis regardless. There is no doubt that the first rule is the best of the four. The question is why it is best.

To answer this question we need to consider the *metadecision* problem in which our options are the above four possible decision rules. The relevant states of the world are, as before, the truth or the falsity of the hypothesis. The matrix for this metaproblem is shown in figure 9.3, where (**A, R**) represents the first rule. Following standard procedures in classical decision theory, the value attached to an outcome is the *expected value* of that outcome given the corresponding state of nature and decision rule. Rule 3, for example, yields an expected value of **1** if **H** is true and **0** if it is false.

The obvious difficulty we now face is that we do not know what is the expected value of adopting Rules 1 or 2 in the case where **H** is *false*. The decision-theoretic approach to the acceptance of hypotheses thus forces us to consider a question that might otherwise go unnoticed. The question will surely go unnoticed if one considers only those aspects of the problem that are determined by logic alone. The negation of **H** does not logically imply whether or not the observations will agree with **H**.

To give a decision-theoretic account of why Rule 1 is correct requires some additional considerations sufficient to determine the missing expected values in the metadecision matrix of figure 9.3. The route suggested by classical decision theory, though not the only logically possible route, is to look for some way of determining the *probability* of observations implied by **H** if **H** is nonetheless false. Let us postpone questions about the nature of these probabilities and how they might be determined in order first to see why some probabilities of this form do provide a solution to the metadecision problem—and thus to the original problem of deciding whether to accept or reject **H**.

It is fairly easy to see that what we need is a *high* probability of the observations being *inconsistent* with **H** if **H** is false. In that case, using Rule 1 results in there being a high probability that we make the correct decision of rejecting **H**. The missing expected value for the rule (**A, R**) is thus close to **1**, while the corresponding expected value for the opposite rule (**R, A**) is near zero. In this case, choice of the intuitively correct rule (**A, R**) could be justified by the minimax principle, since this rule gives a nonzero expected value whether **H** is true or not.

But minimax justifies too much. In a real scientific context, an expected value minimally greater than zero would not be enough. One wants an expected value that is reasonably high. So the operative decision strategy is something like *satisficing*.[12] Then the (**A, R**) rule is the correct choice because it guarantees a sufficiently high expected value whether **H** is in fact true or not. Fixing the value scale as we have, this is equivalent to saying that (**A, R**) is best because it yields a high probability of a correct decision regarding **H** whether **H** is true or not.[13]

There remains the problem of justifying the strategy of satisficing. My view is that satisficing is justified simply on the grounds of being an efficient strategy to follow in a world in which one cannot guarantee correct decisions. This is an empirical claim. I will later reply to those who would at this point raise the specter of a Humean regress.

Nothing in the above decision-theoretic model of acceptance distinguishes empiricist from realist hypotheses. It works equally well for both. If there is a difference, it will have to be found either in the values assigned or in the probabilities which we have thus far left unanalyzed. At this point I think it best to proceed in the context of a new example. The example has been selected because it nicely illustrates the virtues of a realistic image of science.

## 6. The Double Helix

In the fall of 1951, James Watson arrived at the Cavendish Laboratory in Cambridge intent on discovering the chemical structure of DNA. He was hardly starting from scratch. Behind him was Chargaff's determination of the ratios of the four basic nucleotides in DNA and Pauling's discovery of the helical structure of the polypeptide, $\alpha$-keratin. There was also much knowledge of the structure of chemical bonds for various atoms and considerable experience with X-ray diffraction. And Watson had inside information both about Pauling's progress on DNA and about X-ray studies going on at King's College in London.[14]

Following Pauling's example, Watson and Francis Crick began building models of DNA—not theoretical models, but actual *scale* models using bits of wire, metal, and cardboard. One of the beauties of the model-based approach to scientific theories is that this sort of activity can be seen as continuous with the development of theoretical models using language and mathematics. The scale model provides a satisfactory means of characterizing a theoretical model. In this case the theoretical model is an idealized nucleic acid with idealized subunits composed of idealized individual atoms in very specific spatial arrangements. Watson's hypothesis was that actual DNA molecules are similar in composition and structure to the theoretical model. In practice, of course, Watson and Crick referred directly back to the scale model without explicitly invoking the theoretical intermediary.

At the time, the clearest sort of evidence for such structural hypotheses was provided by the techniques of X-ray diffraction. The diffraction patterns and relative intensities produced on photographic plates tend to be quite sensitive to the structure of the material producing the pattern. On the other hand, no one at the time knew enough about the X-ray patterns produced by nucleic acids to be able simply to read off the structure of the material from an X-ray photograph. If that had been possible, the structure would have been known long before Watson arrived on the scene. Once the double helical structure had been proposed, workers at King's College quickly reexamined photographs already in their possession. The results agreed substantially with those predicted by the Watson-Crick model. Papers discussing this evidence appeared in the same issue of *Nature* as the original Watson and Crick paper.

Because of the complexities of interpreting X-ray photographs and uncertainty about the role played by photographs already in Watson's possession, the confirmation in this case was not as dramatic as in some other recent cases, e.g., the confirmation of continental drift or of big-bang models of the universe. But the DNA case is clear enough and more relevant to an examination of the relative virtues of empiricism and realism. I will proceed first to argue the minimal thesis that a realistic hypothesis can after all be justified.

The realistic hypothesis, of course, is that the molecular structure of DNA is quite similar in the relevant respects to the proposed model. The evidence is the particular pattern of light and dark spots observed on X-ray photographs. The decision rule is to accept the hypothesis if the observed pattern matches that predicted by the model and reject it if the pattern fails to match. Our choice of this decision rule is justified by the principle of satisficing if we can show that the *probability* of getting the pattern is *high* if **H** is true and *low* if **H** is false.

Enough was known about the patterns produced by various molecular structures and about X-ray techniques to justify the *conditional* judgment that, if the hypothesis were correct, the specified pattern would very likely result. But the same body of background knowledge also justified the converse judgment. If the real structure were not very similar to the model, this particular pattern would be unlikely to result. One does not have to know just which patterns go with which structures to know that importantly different structures rarely yield similar patterns. Moreover, the theory of X-ray diffraction substantiates this judgment—that is, it tells us why the observed pattern should be quite sensitive to differences in structure. Thus, even if we set a quite high satisfaction level, the standard decision rule is justified—and so, therefore, is the realistic hypothesis.[15]

It might be objected that we were able to justify the realistic hypothesis only because we assumed that some *other* realistic, indeed, *modal,* hypotheses were already justified. In particular, we appealed to our background knowledge to determine what *would* be the likely result if **H** *were* true and if it *were* false. This is correct. But does this circumstance constitute an objection to realism? We have not literally begged the question by assuming that the hypothesis under investigation is justified. We have only assumed that some *other* realistic hypotheses were previously justified.

To argue that the above procedure begs the question is, I think, to misconceive the problem. Our task was to show how realistically interpreted hypotheses are justified. That we have done. What we have not done is show how a realistic hypothesis could be justified from a background consisting only of observed facts—perhaps together with empirical generalizations. This is a very different, and peculiarly philosophical, problem. It might be confused with the old problem of induction, in which case it is simply impossible to solve. Suitably interpreted, however, it can be solved, as I shall shortly attempt to show.

Right now let us see how the empiricist version of the Watson-Crick hypothesis fares in our decision-theoretic framework. The hypothesis is that the model is empirically adequate to the phenomena of X-ray photographs. From this hypothesis it follows immediately that the observed patterns will match. To justify the standard decision rule, however, we need to know the probability of a match if the proposed model is *not* adequate. Now, for the empiricist, our background knowledge consists of other models previously justified as being empirically adequate. In particular, the empiricist knows that models of molecular structure differing substantially from empirically adequate models are rarely themselves empirically adequate. Employing this knowledge, the empiricist concludes that the predicted match is unlikely if **H** is false. This shows that the decision rule is satisfactory and that the observation of the predicted pattern does justify the hypothesis.

What the empiricist cannot do, however, is appeal to knowledge of the internal causal structure of nucleic acids and their interactions with X rays in order to bolster the judgment that the predicted pattern is indeed unlikely if **H** is false. The empiricist's knowledge is exhausted by the claim that these other models are empirically adequate. Unjustified claims about causal structure cannot provide justification for their consequences. Thus, the realist claim that the decision rule is sufficiently justified may itself be better justified than the corresponding empiricist claim. And surely the acceptance of a hypothesis using a better-justified decision rule is itself better justified. So the realist hypothesis might after all be *better* justified than the empiricist hypothesis.[16] I say only that the realist hypothesis "might" be better justified because we have not yet taken account of the logical principle that safety is inversely related to strength. But at least we have shown the possibility that in science, as in diplomacy, there may be security in strength.[17] To pursue this possibility, we must consider two more cases.

First, let us return to the philosophical problem of whether a realistic hypothesis could be justified assuming only a background of empirically interpreted hypotheses. In fact, it is fairly easy. All we need do is substitute the realistic Watson-Crick hypothesis for the empiricist version in the empiricist's decision problems. The decision rule is justified by appeal only to empiricist hypotheses. In the end we accept the realist hypothesis. We reach realist goals using empiricist means. The reverse strategy, pursuing empiricist goals using realist means, is also possible. That is, we might justify the empiricist version of the Watson-Crick hypothesis using realistic background knowledge to justify our decision rule. Though philosophically unmotivated, the possibility of pursuing such a strategy prepares us to face the issue of safety versus strength.

The decision-theoretic approach to acceptance has introduced a new factor in addition to the principle that safety is inversely related to strength. For a fixed hypothesis, the amount of justification is greater if one begins with a better-justified decision rule. Or, to put it in more familiar terms, the more *severe* the test, the better the justification—where the severity of a test is proportional to the *improbability* of the implied result if the hypothesis is false.[18] And we have argued that using realistic background knowledge provides the basis for a stronger test. Using these two principles, then, we obtain the following partial ordering of justifications.

The *strongest* justification is obtained if we judge the empiricist hypothesis using realistic background information to justify the decision rule. The realist hypothesis, being logically stronger, would be less well justified on the same background knowledge. At the other end, the *weakest* justification is obtained if we judge the realist hypothesis using empiricist background information. The empiricist hypothesis is better justified on the same empiricist background knowledge. On the other hand, keeping the hypothesis fixed and varying the background knowledge yields the result that the empiricist hypothesis is better justified on realistic background knowledge than it is on empiricist background knowledge. Similarly, the realist hypothesis is also better justified with a realistic rather than an empiricist background.

All these relationships, however, fail to determine whether the realist hypothesis on realistic background knowledge is better justified than the empiricist hypothesis with an empiricist background. We don't know the rate of exchange between justification due to severity of test and to logical strength of the hypothesis. Nor does it seem that any such rate could be determined without a logical measure of content and confirmation, something van Fraassen and I both reject. So the opposition between constructive empiricism and constructive realism cannot be decided merely by appeal to considerations regarding safety and strength.

The decision-theoretic framework suggests another sort of appeal—*value*. The realist values realistic hypotheses over empiricist hypotheses, regardless of the background knowledge. This value difference may compensate for the difference in logical strength. The appeal to value, however, obviously cannot decide the issue, since the empiricist has contrary values. But it does put the issue in better focus. The question for van Fraassen is why he prefers empiricist hypotheses. It cannot simply be a matter of epistemic caution, since he values actual empiricism over extreme empiricism. He is willing to take some risks. Why is he unwilling to take a few more? Nor is it sufficient to point to the specter of metaphysics. The physical modalities he would regard as metaphysical are for others the essence of theoretical science.

Perhaps, after all, he simply can't let go of the old Hume in our philosophical tradition. Since the old Hume dies hard, it may help to ease the grief to consider a Darwinian tale. Like van Fraassen, I am a Darwinian. Indeed, I am more so, since I believe not only in the facts of evolution evidenced by the fossil record but also in the reality of the mechanisms—modeled in part by the genetic theory of natural selection—that produced this record.

My tale begins with human beings and human culture evolving together over many centuries. This evolution could not have been successful if the culture did not both generate and transmit a fair amount of fairly reliable, though perhaps mundane, knowledge about the natural world. This knowledge, I claim, must have included some understanding of the causal mechanisms underlying the phenomena. The development of science, solidified in the west in the seventeenth century, consisted both in acquiring the *goal* of learning more details about the deep structure of the natural world and in devising *means* appropriate to this goal. The stock of knowledge developed over the past three hundred years provided the background for the current highly accelerated growth of science.

Filling in the details of this story would not, of course, provide a philosophical *justification* of induction. The tale does, however, provide the outlines of an *explanation* of induction. To provide more is neither necessary nor possible.[19]

## 7. The Theory of Science

One way of understanding *The Scientific Image* is as an attempt to construct a *theory of science*.[20] Moreover, in the terminology of some recent sociologists of science, the theory is fully *reflexive*. That is, it has itself as an instance.[21] From this standpoint, we see van Fraassen constructing a model of science and attempting to show that the model is empirically adequate. His arguments, however, proceed at a high level of abstraction. He is concerned to show, for example, that the goal of empirical adequacy is sufficient to account for philosophical theses about methodology, explanation, causation, and probability. Only occasionally, as in the Millikan example, does he argue the case at the level of scientific practice. But it is at this level that empiricism as a theory of science stands or falls. So we must ask, is constructive empiricism empirically adequate?

My introduction of an example from molecular biology was partly motivated by this question. It is very difficult, I think, to save the phenomena of molecular biology, *as a scientific enterprise,* using the empiricist model. The major figures in the 1950s, Pauling, Watson, Crick, et al., could not be described as having the goal of merely accounting for the phenomena of X-ray diffraction and chemical reactions. Watson clearly had little interest in X-ray pictures except as a means to the end of discovering the structure of DNA. Rather, much time in Crick's laboratory was spent measuring the angles between the metallic representations of nucleic acids in their scale model to see if the model fit the known bonding angles of various atoms. These are not the actions of people striving merely to account for spots on X-ray photographs.

Nor is the picture much changed if we move from talking about the goals of individuals to considering the goals of the science as a whole. Reading the history of molecular biology and excerpts from papers and letters written at the time, it is difficult to find any evidence of an overriding concern with saving the phenomena. The whole profession acted as if it were after the real molecular structure of real molecules.

Van Fraassen says that scientists live in a life-world created by their models. They talk *as if* they take their models to be descriptive of reality. To adapt what he says about Millikan, "it may be natural to use the terminology of discovery" to describe the achievements of the early molecular biologists, but "the accurate way to describe it is that [they were] writing theory by means of [their] experimental apparatus." We should, van Fraassen urges, take "a purely functional view" of the relation between theory and experiment. "Experimentation is the continuation of theory construction by other means" (1980, 74–77).

Van Fraassen's willingness radically to reinterpret how scientists talk and act raises serious questions about the phenomena which a theory of science is to save. Of course, an individual's words and even deeds may belie his real goals. But are we to be free radically to reinterpret the words and deeds of a whole scientific community? If so, do there not cease to be any phenomena that could count against the adequacy of any theory of science? If this is so, then the theory of science ceases to be a scientific theory even on empiricist grounds. My worry, then, is that van Fraassen's constructive empiricism can avoid empirical inadequacy only by forfeiting any claim to being a genuine theory of science.[22]

Putting aside these global worries, let me conclude on a more harmonious note. The model-based account of theories permits a theory of science which avoids one of the major methodological defects of much social science—the demand for *universality*. If the empiricist model does not fit molecular biology, that does not mean it is worthless as part of a theory of science. The question is whether there are *any* major sciences, or long periods in the life of some major sciences, that fit the empiricist model. It seems hard to deny that there are. Greek astronomy, thermodynamics in the late nineteenth century, and quantum theory in the twentieth century are obvious candidates. It may be more than coincidence that quantum physics is the science van Fraassen knows best. On the other hand, many contemporary sciences, including chemistry, molecular biology, and geology, seem decidedly realistic.

Of course, many disagreements remain. For example, is empiricism or realism the dominant mode in contemporary science? I would guess that realism is dominant. Empiricism is the mode of sciences lacking a solid theoretical tradition, e.g., many of the social sciences, or of sciences whose models are difficult to understand in spite of much empirical success, e.g., quantum theory. Another issue is whether having realistic goals tends to make scientists, or even scientific fields, more successful in discovering new results. Again I would guess that realism is scientifically more fruitful. Though difficult to resolve, these issues, happily, are more empirical than theoretical.

# TEN

# The Feminism Question in the Philosophy of Science

## 1. Introduction

The title of this chapter is a reflection of Sandra Harding's *The Science Question in Feminism* (1986). Her science question in feminism is this: Feminist claims of masculine bias in science are often themselves based on scientific studies, particularly the findings of various social sciences. But if the claims or methods of science are in general as suspect as many feminists claim, then appeals to scientific findings to support charges of bias are undercut. In short, is it possible simultaneously to appeal to the authority of science while issuing general challenges to that same authority?

My feminism question in the philosophy of science is this: To what extent is it possible to incorporate feminist claims about science within the philosophy of science? Are feminist claims about science compatible with a philosophy of science that rejects relativism? Are they compatible with a philosophy of science that embraces realism? In short, how seriously should philosophers of science, in general, take the claims of feminists that the philosophy of science should incorporate feminist claims about science? The answer to my question, of course, depends both on what feminist claims one considers and on one's conception of the philosophy of science.

From the standpoint of the philosophy of science, the most significant claim of feminist scholars is that the very *content* of accepted theory in many areas of science reveals the gender bias of the mostly male scientists who created it. Moreover, the theories in question came to be accepted through the application of accepted methodological practices. So the sciences and scientists involved cannot be written off as obviously biased or otherwise marginal. Thus, gender bias in the content of accepted science is both possible and, in some cases, actual.

## 2. Case Studies

An appropriate starting point for an examination of feminist critiques of science is with the many case studies of actual scientific research purporting to demonstrate masculine bias in the results of what had been regarded as clear cases of acceptable scientific practice. Investigating such cases, however, is much more difficult than one might think. Before explaining why, I will provide a rough taxonomy of cases and mention a few examples.

The most convincing cases are those in which the subject matter of the science consists of either real human beings or higher mammals, and the theories in question focus on aspects of life in which sex or gender is obviously a variable. This includes parts of many sciences such as anthropology, sociology, ethology, and primate evolution. Standard examples of these sorts of cases include theories of human evolution based on a model of "man the hunter." According to these theories, the evolution from higher primates to humans was driven by selective forces operating in small groups of male hunters. The use of tools, the development of language, and particularly human forms of social organization, have all been claimed to have evolved in the context of hunting by males. This theory has been the standard theory in many fields for several generations. This approach was not seriously challenged until women entered these fields in more than token numbers and began developing an alternative model of "woman the gatherer." These women have argued that gathering and elementary agriculture likewise require complex skills, social organization, communication, and the development of basic tools. And, they argue, the evidence for this theory is at least as good as that for the standard "man the hunter" paradigm. The lesson drawn is that the "man the hunter" account was the accepted theory for so long at least in part because it was developed and sustained by scientific communities dominated by men with masculine values and experiences. Developing a plausible rival required women with female values and experiences.[1] The investigations of Longino and Doell (Longino 1990, chs. 6 and 7) into theories of the biological origin of sex differences in humans provide another outstanding example of this type of case.

A second category consists of cases in which the subjects are humans or primates, but the theories are not directly about obviously sexed or gendered aspects of their lives. Here a good example comes from the field of psychological and moral development. The standard theories for most of the twentieth century were those developed by Freud, Erikson, and Kohlberg. These theories purported to be theories of "human development" but were in fact based primarily on studies of boys and men. When studies of girls and women were made, observed differences were treated as "deviations" from the established norm, or even as evidence of failure by girls to reach the higher stages of development. A contrary view emerged in the 1970s through the work of female psychologists such as Carol Gilligan as reported in her now classic book *In a Different Voice* (1982). Gilligan studied moral development in both men and women, but concentrated on women. Her conclusion was that women are neither deviant nor lagging in their moral development, just different.[2] The lesson is the same as that from the "man the hunter" model.

A third category of cases involves living but non-mammalian subjects, and theories in which sex is not a salient variable. A possible example here is Barbara McClintock's work on genetic transposition as interpreted by Evelyn Fox Keller in her 1983 book *A Feeling for the Organism*. Keller argued that McClintock approached her subject with values and interests that were connected with the fact that she was not a man in a profession dominated by men. McClintock, Keller claimed, had an appreciation for complexity, diversity, and individuality, and an interest in functional organization and development, which was at variance with the desire for simple mechanical structures that motivated most of her male colleagues. That, according to Keller, explains both why McClintock was able to make the discoveries she did, and why her mostly male colleagues failed for so long to understand or appreciate what she had done.[3]

The fourth and most difficult category for the feminist critique involves nonliving subjects, and theories that obviously do not explicitly incorporate sex or gender as a relevant variable. This includes sciences from molecular biology to high energy physics. Here Keller (1985, 1992, 1995) and a few others have argued that the influence of gender can be seen in the metaphors that, they claim, both motivate and give meaning to the theories that are generally accepted. DNA, for example, is thought of as a kind of genetic control center issuing orders along a hierarchical chain of command—a clearly male military or corporate metaphor.

For any of these cases to be effective as a critique of science, one must maintain *both* that they exhibit a clear masculine bias *and* that they nevertheless constitute examples of acceptable scientific practice. To dismiss the cases, therefore, one can argue either that the case for masculine bias is not sufficiently substantiated, or that bias does exist, but the cases are not acceptable science. The power of the anti-feminist position lies in the fact that one can use the argument for gender bias as itself grounds for concluding that the case is one of bad science, thus undercutting the feminist critique. And this strategy is likely to be most successful in the examples where the prima facie case for masculine bias seems strongest. Suspicion of the scientific credibility of such "soft" sciences as anthropology and cognitive development long antedated feminist critiques of theories in these fields.

I believe that a credible case for the feminist position has been made in at least some of these examples, but this claim can only be substantiated by a detailed examination of the cases themselves. So, rather than engage the debate at this level, I will shift my attention to the question whether it is *theoretically possible* that the feminist conclusion is correct. Could there be gender bias in what by all other criteria must count as good science? There is a rhetorical as well as a theoretical reason for raising this question. Many philosophers, including philosophers of science, simply do not regard it as theoretically possible that the feminist critique could be correct. For these philosophers, looking carefully at the cases is merely an academic exercise. To be convinced, therefore, that it is worth even considering the implications of the feminist critique for the philosophy of science, one must first be convinced that it is at least theoretically possible that the critique is correct. That is what I hope to do here—make a convincing case that it is theoretically possible.

## 3. Some Sources of the Anti-Feminist Position

I will now consider several sources of the presumption that the feminist position is theoretically impossible. If it can be shown that this presumption rests on inadequate foundations, that would undercut the anti-feminist position.

One source is the Enlightenment ideal of science. The cornerstone of the Enlightenment ideal is the view that the ability to acquire genuine knowledge of the world is independent of personal virtue or social position. Popes and Bishops, Kings and Knights, have no special access to genuine knowledge. What matters is the correct employment of natural reason, and that is, in principle, within the grasp of any normal person. The irrelevance of gender was presumed, although too often because women were deemed not capable of exercising the powers of natural reason. In the present day philosophical canon, most of the thinkers between Descartes and Kant held an Enlightenment picture of science, even if, like Descartes, they were precursors rather than participants in the Enlightenment as such. To a large extent, much of contemporary philosophy simply presupposes this Enlightenment ideal. And that at least partly explains why so many contemporary philosophers and philosophers of science find it simply impossible that gender might matter for what counts as legitimate scientific knowledge.

Feminists, not surprisingly, tend to take a dim view of the Enlightenment. I would urge a middle ground, insisting that the Enlightenment was an genuine advance over what came before, but recognizing that its presumption of the gender neutrality of human reason was *merely* a presumption, and not based on any firm grounds, particularly not the sorts of empirical investigations now common in the cognitive and social sciences. But I do not want to dwell on the Enlightenment. There are sources much closer to our own time for the view that the feminist critique could not possibly be correct.

The current configuration of views within philosophy of science in the United States derives mainly from European sources transmitted by refugees displaced by World War II. For the most part, these influential refugees were German speaking members of a group advocating a scientific philosophy, a "Wissenschaftliche Weltauffassung." These thinkers were repelled by the various neo-Kantian idealisms then dominant within German philosophy, and in German intellectual life generally. And they were simultaneously inspired by the new physics associated above all others with the work of Einstein.

In a nutshell, the position of the scientific philosophers was that to understand the nature of fundamental categories like space and time, one should look to Einstein's relativity theory, not to the *a priori* theorizing of neo-Kantian philosophers. Similarly, to understand the nature of causality, one should look to the new quantum theory. Their program was a radical program, a program to *replace* much of philosophy as it was generally practiced in Germany with a new scientific philosophy. It is thus not surprising that none of these philosophers occupied positions of great influence, whether intellectual or institutional, within the German speaking philosophical world.

The most prominent at the time was Moritz Schlick, Professor of Philosophy at the University of Vienna. Schlick was not really part of the Viennese philosophical establishment, but he provided both philosophical inspiration and institutional support for the Vienna Circle. It was he who, in 1926, brought the young Rudolf Carnap to Vienna as an instructor in philosophy. And it was Carnap who became the intellectual leader of the Vienna Circle, a heterogeneous group of mathematicians, natural scientists, social scientists, and scientifically trained philosophers like himself.

Most prominent among the scientific philosophers outside of Vienna was Hans Reichenbach in Berlin. While a student of physics and mathematics in the teens, Reichenbach was active in socialist student movements. That ended when he began teaching science and mathematics in various Technische Hochschulen. He also began publishing logical-philosophical analyses of Einstein's theory of relativity. In 1927, Einstein, together with Planck and von Laue, arranged for Reichenbach to be offered an untenured position in the physics department at the University of Berlin. The philosophers in Berlin voted not to admit Reichenbach as a member of their department, but Einstein, at least initially, welcomed his help in carrying on his own intellectual battles with the neo-Kantians over the nature of space, time, and causality. Reichenbach relished the role.

With the imposition of the Nazi racial laws in the spring of 1933, Reichenbach, along with hundreds of other German professors, was dismissed from his post. Einstein, having resigned from abroad, found safe haven at the newly created Institute for Advanced Study in Princeton. Reichenbach was among fifty or so former German professors who accepted five-year contracts at the University of Istanbul. Even before his call to Berlin, Reichenbach had been exploring the possibility of emigrating to the United States. Now he resumed these efforts in earnest. As part of his plan to find a position in the United States, he put aside his technical work both on relativity and on the theory of probability, and began writing, in English, a general work on scientific epistemology. That work, *Experience and Prediction,* was completed in 1937 and published by the University of Chicago Press in 1938—the year Reichenbach began his tenure at UCLA.

In very first section of that book, titled "The Three Tasks of Epistemology," Reichenbach introduces his distinction between "the context of discovery" and "the context of justification," remarking that "epistemology is only occupied in constructing the context of justification" (7). The introduction of the distinction is not the conclusion of any argument. It is a precondition for the analysis to follow. In fact, this distinction, though of course not in these words, had existed in German philosophy for half a century. But this seems to be the first time it appeared in Reichenbach's writings. It reappears only once in *Experience and Prediction,* near the end of the final chapter on probability and induction, where, explicitly referring to the example of Einstein, he writes that induction "is nothing but a [logical] relation of a theory to facts, *independent of the man who found the theory*" (p. 382, emphasis added).

I would speculate that when Reichenbach introduced the distinction between discovery and justification he was explicitly motivated by the case of Einstein, whose views were vilified in the Nazi press not because of any lack of a proper logical relation between his theories and the facts, but simply because of a personal fact about the man with whom those theories originated—he was a Jew. Reichenbach's own personal situation differed from Einstein's mainly in that his accomplishments, and consequently his reputation, were less exalted.[4]

One can now see a clear connection between contemporary feminist critiques of science and Reichenbach's use of the distinction between discovery and justification. Reichenbach, I believe, made it a precondition for doing scientific epistemology that the very notion of "Jewish science" be philosophically inadmissible. The Nazi racial laws were not only a crime against humanity, they were a crime against philosophical principle. The feminist notion of "masculine science," or any sort of gendered science, is not in principle any different. It makes the epistemological status of a scientific theory dependent on facts about the scientists themselves, as historical persons, quite apart from internal, logical, relations between fact and theory.

Even if I am mistaken about the personal motivation behind Reichenbach's use of a then well-known distinction in his first general epistemological work, there is no doubt that his understanding of the distinction rules out the relevance of gender to any philosophically correct understanding of legitimate scientific knowledge. Moreover, this understanding of the task of scientific epistemology was shared by most of the European scientific philosophers. And it was these philosophers who came to dominate philosophical thought about science in the United States during the postwar period.

One might object that this is all just so much history of the philosophy of science. Where are the arguments? I hope it is clear that this response begs the question at issue. The validity of the discovery-justification distinction was not established by argument. It was, as is clear in Reichenbach's book, part of the initial statement of the task of a scientific epistemology. It is part of that conception of scientific epistemology that gender or other cultural factors cannot possibly play any role in establishing the legitimacy of scientific claims. My "argument" has been that it is to a large extent due to the legacy of those whose conception of the philosophy of science was formed in the war against Nazi power and ideology that the idea of gendered science still seems to many simply impossible.

The point of my historical remarks can be put more sharply. The insistence on the irrelevance of origins which has characterized Logical Empiricism in America is refuted by the history of that movement itself. The prominence of many doctrines, like the discovery-justification distinction, was not the result of argument, but an assumption forming the conceptual context within which arguments were formulated. To understand fully why those doctrines were held, one must inquire into the historical origins of their role in that movement. Indeed, it is a revealing irony that later criticisms of the discovery-justification distinction focused almost exclusively on its validity or usefulness, not on its origins.[5]

## 4. The Possibility of Gender Bias in Postpositivist Philosophy of Science

The contemporary feminist movement in America has its own roots in the civil rights movement and the antiwar movement of the 1960s. That was a different war, a different generation, and a different set of political circumstances. The major influence on the philosophy of science of that decade was Thomas Kuhn's *Structure of Scientific Revolutions*. Kuhn clearly did not set out to become a hero of the 1960s cultural revolution. Nor could one who wrote so unselfconsciously about "the man of science" have been promoting a feminist agenda. Yet his work has, I think correctly, e.g., by Keller (1985), been seen as providing support for the possibility of gendered science.

In Kuhn's book, the distinction between "the context of discovery" and "the context of justification," in just those words, appears again in the very first chapter. Here, however, Kuhn himself remarks that the distinction seems not to have been the result of any investigation into the nature of science. Rather, he claims, it was part of a framework within which the study of science had been carried out. He makes it clear that his own inquiry does not presuppose any such distinction. And, indeed, Kuhn's own theory of science, with its emphasis on the role of individual judgment exercised by scientists in communities, yields nothing that would rule out the possible influence of gender on the eventual beliefs of a typical scientific community.

In the philosophical profession at large, it is widely believed that Kuhn was part of a historical turn in the philosophy of science which superseded Logical Empiricism. That belief is mistaken on at least two counts. First, the historical tradition within the philosophy of science did not supersede Logical Empiricism. It was, rather, a rival philosophical tradition which emerged around 1960 and was in part stimulated by Kuhn's work. Logical Empiricism continued to evolve both in terms of the study of particular scientific theories and in terms of general methodological inquiries. Both sorts of developments are exemplified, for example, in the works of Bas van Fraassen (1980, 1989, 1991). Second, Kuhn himself was only marginally a part of the historical tradition within the philosophy of science. Most of the philosophers of science associated with that tradition, including Paul Feyerabend, N. R. Hanson, Imre Lakatos, Larry Laudan, Ernan McMullin, Dudley Shapere, and Stephen Toulmin, shared Kuhn's rejection of Logical Empiricism. And they agreed with his focus on scientific development as a central notion for the study of science. But, for the most part, they also shared a rejection of Kuhn's own theory of science.

With the obvious exception of Feyerabend, these historically oriented philosophers of science sought not to reject the logical empiricist idea of an objective connection between data and theory, but to *replace* the idea of a *logical connection* between data and theory with that of *rational progress* within a research tradition. This shift is clearest in the case of Lakatos. For Lakatos, a research program is progressive to the extent that it generates successful novel predictions yielding new confirmed empirical content. There appears to be no room in this definition for any influence from cultural variables such as gender. I will now argue that the apparent impossibility of gender bias in postpositivist philosophical theories of rational progress is only apparent. It is possible even on Lakatos's hard-line account.

One of the many lessons Kuhn claimed to have learned from his study of the history of science was that scientists rarely abandon a research tradition unless they first can at least imagine a promising alternative. Both Lakatos and Laudan explicitly adopted this idea, arguing that the evaluation of a research tradition is not based on a two-place relationship between data and a theory, but on a three-place relationship between data and at least two rival research traditions.

There cannot be many examples in the history of science where the existing rival research programs exhaust all the logical possibilities. So it is typically possible that the theories making up the existing rival research programs are in fact all false. Nevertheless, as Kuhn argued, and almost everyone else agreed, it is rare to have a scientific field in which there is no clearly favored research program. There is typically an establishment position. It follows that, at any particular time, which research program is most progressive by any proposed criteria depends on which of the logically possible research programs are among the actually existing rival programs. Against other logically possible rivals, the current favorite might not have fared so well. Moreover, not only Lakatos and Laudan, but most others as well, retain a distinction between discovery and justification to the extent that their accounts of rational progress place few if any constraints on how a possible research program comes to be an active contender. There is little to rule out this process being driven by gender bias or any other cultural value.

So, for any leading research program, it is possible that its position as the current leading contender is in part a result of gender or some other cultural bias. If these biases had been different, other programs might have been considered, and a different program might have turned out to be comparatively more progressive at the time in question. In short, the fact that a given program is judged normatively most progressive by stated criteria might possibly be due, at least in part, to the operation of gender biases in the overall process of scientific inquiry. And that is enough to establish the possibility that the feminist critique is correct in at least some cases.

## 5. A Popperian Response

My earlier survey of leading scientific philosophers omitted any mention of Karl Popper. That was deliberate, because, as I see it, Popper had little influence on what became Logical Empiricism, particularly in America, until after publication of the 1959 English edition of his 1935 monograph *Logik der Forschung* under the even more misleading title *The Logic of Scientific Discovery*. Despite his own claims that it was he who killed positivism (1974), the accidental fact that the English edition of Popper's book appeared shortly before Kuhn's put him in a position to become a primary defender of the positivist faith against the Kuhnian heresy.

The titles of Popper's book are misleading because, on his account of science, there is no such thing as a "logic" of research or of scientific discovery. The main role for logic in science is the use of *modus tollens* in the refutation of a universal generalization by a statement describing a negative instance. This form of inference requires no reference to alternative hypotheses. So, apart from questions about how one establishes the truth of the required singular "observation statement," this form of inference would seem to be immune to gender or any other cultural influences. Popper's work thus shows that it is possible to construct a theory of science which maintains a strong enough distinction between the contexts of discovery and justification to eliminate the possibility of gender bias. But it also shows how very difficult it is to construct a *good* theory of science that fulfills this requirement. No one better exhibited the shortcomings, not to say the utter implausibility, of Popper's theory of science than his successor, Imre Lakatos—and Lakatos borrowed heavily from Kuhn. It should be noted that the approaches to scientific justification taken by both Carnap and Reichenbach would, if successful, also eliminate any possibility for gender or other cultural biases. For both, theory evaluation is not comparative, at least not in any obvious way. I will not elaborate this point further because these approaches have few defenders today.

The successor to Carnap's conception of inductive logic is a subjective probability logic, as championed, for example, by Carnap's associate, Richard Jeffrey (1965). Theories of subjective probability, however, place only minimal constraints on how an individual assigns initial probabilities to any theory. This leaves lots of room for individual scientists to assign high initial probabilities to theories reflecting their own particular gender biases. The best the probabilistic approach can offer is proof of the diminishing influence of the initial probability assignment in the face of increasing observational evidence. But there is no way of knowing, in this framework, how much the probability assigned a particular theory at any given time might be the product of some form of bias, including gender bias. That leaves feminist critics as much room as they need.

In sum, there is little in current philosophical theories of science that supports the widespread opinion that gender bias is impossible within the legitimate practice of science. That opinion seems mainly the product of a traditional adherence to an Enlightenment ideal of science strongly reinforced by the historical origins of twentieth-century scientific philosophy in Europe and its rebirth as Logical Empiricism in America. As disquieting as it may seem to many, we shall have to learn to live with the empirical possibility of "Jewish science." That is, for any particular scientific theory, it must be an *empirical* question whether its acceptance as the best available account of nature might be due at least in part to its having been created and developed by Jewish scientists rather than scientists embodying some other religious tradition. In another cultural context in which science as we know it is generally practiced, some other theory might now be the accepted theory. Whether or not this is true for any *particular* theory can only be determined empirically by looking in detail at the history of how that theory achieved its present status. The irrelevance of religious origins cannot be guaranteed *a priori*. The same holds for gender.

## 6. Perspectival Realism

In countenancing the relevance of cultural forces in the acceptance of scientific theories, have we not moved too far in the direction of *relativism?* In particular, is this position compatible with a reasonable scientific realism? I think it is, but the issue is complex. If we suppose that the world is organized in a way that might be mirrored in a humanly constructible linguistic system, then there is indeed a problem. For then realism seems to require that we could have reason to believe that our theories are literally *true* of the world. The objects in the world are grouped as our theories say they are and behave as our theories say they should behave. If, however, what we *take to be true* of the world is influenced by cultural factors, there is no reason to think that this influence would promote the development of actually true theories and considerable reason to suspect that it would do just the opposite. That sounds like relativism, not realism.

Radical though it may seem, I think the solution to this problem is to reject the usefulness of the notion of *truth* in understanding scientific realism. I do not mean that we cannot use an everyday notion of truth, as when asserting that it is indeed true that the earth is round. Here truth may be understood as no more that a device for affirmation. Rather, it is the *analysis* of truth developed in the foundations of logic and mathematics, and used in formal semantics, that we should reject in our attempts to understand modern science. But if we reject the standard analyses of truth and reference, what resources have we left with which even to formulate claims of realism for science? The answer is that the notion of linguistic truth is but one form of the more general notion of *representation*. What realism requires is only that our theories well *represent* the world, not that they be true in some technical sense. So we need a notion of representation for science that does not rest on the usual analyses of truth for linguistic entities. What might that be?

A first step is to reject the analysis of scientific theories as sets of statements in favor of a model-based account which makes nonlinguistic models the main vehicles for representing the world, and places language in a supporting role.[6] We may, of course, use language to characterize our models, and what we say of the models is true. But this is merely the truth of definition, and requires little analysis. The important representational relationship is something like *fit* between a model and the world. Unlike truth, fit is a more qualitative relationship, as clothes may be said to fit a person more or less well. Of course we can say it is true that the clothes fit, but this is again merely the everyday use of the notion of truth.

Here I can offer no general analysis of the notion of fit, only a further analogy—maps. There are many different kinds of maps: road maps, topological maps, subway maps, plat maps, etc. And it can hardly be denied that maps do genuinely represent at least some aspects of the world. How else can we explain their usefulness in finding one's way in otherwise unfamiliar territory? Moreover, the idea of mapping the world has long been present in science. There were star charts before there were world atlases, and scientists around the world are now busy "mapping" the human genome. Maps have many of the representational virtues we need for understanding how scientists represent the world. There is no such thing as a universal map. Neither does it make sense to question whether a map is true or false. The representational virtues of maps are different. A map may, for example, be more or less accurate, more or less detailed, of smaller or larger scale. Maps require a large background of human convention for their production and use. Without such they are no more than lines on paper. Nevertheless, maps do manage to correspond in various ways with the real world.

Since no map can include *every* feature of the terrain to be mapped, what determines which features are to be mapped, and to what degree of accuracy? Obviously these specifications cannot be read off the terrain itself. They must be imposed by the mapmakers. Presumably which set of specifications gets imposed is a function of the *interests* of the intended users of the maps.

Among cartographers, those whose job it is to make maps, it is assumed that constructing a map requires a prior selection of features to be mapped. Another aspect of mapmaking emphasized by cartographers is *scale,* particularly for linear dimensions. How many units of length in the actual terrain are represented by one unit on the map? These two aspects of mapmaking, feature selection and scale, are related. The greater the scale the more features that can be represented. The required tradeoffs again typically would reflect the interests of the intended users.

It is not stretching an analogy too far to say that the selection of scale and of features to be mapped determines the *perspective* from which a particular map represents the intended terrain. Photographs taken from different locations provide more literal examples of different perspectives on a terrain or a building. In any case, given a perspective in this sense, it is an empirical question whether a particular map successfully represents the intended terrain. If it does, we can reasonably claim a form of *realism* for the relationship between the map and the terrain mapped. I call this form of realism *perspectival realism*.[7]

Standard analyses of reference and truth suggest a metaphysics in which the domain of interest consists of discrete objects grouped into sets defined by necessary and sufficient conditions. Likewise, there is a metaphysics suggested by perspectival realism. Rather than thinking of the world as packaged into sets of objects sharing definite properties, perspectival realism presents it as highly complex and exhibiting many qualities that at least appear to vary continuously. One might then construct maps that depict this world from various perspectives. In such a world, even a fairly successful realistic science might well contain individual concepts and relationships inspired by various cultural interests. It is possible, therefore, that our currently accepted scientific theories embody cultural values and nevertheless possess many genuinely representational virtues.

## 7. Feminist Realism

There is an unfortunate mismatch in terminology between feminist and general philosophers of science. Within the philosophy of science generally, the distinction between empiricists and realists concerns the sort of epistemic commitment one has toward "unobservable" or "theoretical" entities and properties. Empiricists would restrict our commitments to the observable phenomena; realists make no such restrictions. "Feminist empiricist," on the other hand, characterizes someone who thinks some theories may embody gender biases, but also thinks such biases can be detected using standard scientific methods. Moreover, better theories, which may embody other biases, can be proposed and validated. Feminist empiricism, therefore, is neutral regarding the general debate between empiricists and realists. Of course a feminist empiricist might also be an empiricist in the more general sense, but that would be an *additional* commitment beyond feminist empiricism. More significantly for my purposes, a feminist empiricist could be a *realist* in the more general sense. Thus, feminist realism is not inherently an incoherent doctrine. Given current usage, it turns out, misleadingly, to be a special case of feminist empiricism.

To my knowledge, no feminist philosopher of science has claimed to be a feminist realist.[8] I expect this is because realists have often claimed to know *the truth* about many things, or at least to be *rationally justified* in claiming such knowledge. Feminists, quite naturally, are suspicious of any such claims. From my point of view, this suspicion presupposes the mistaken view that realism must be understood in terms of truth in the standard philosophical sense. Abandoning this presupposition, one is free to adopt a perspectival account of realism which is far more congenial to the interests of feminists. Moreover, adopting perspectival realism does not commit one to any form of special scientific rationality. Perspectival realism is perfectly compatible with a thoroughgoing naturalism which appeals only to the naturally evolved cognitive capacities of human agents together with their historically developed cultural artifacts. It is to be expected that such agents would typically project their cultural values, including gender values, into the models they develop to explain phenomena in the world. And some of these models could be expected to end up as part of established science. That is just what feminist philosophers of science have been claiming all along.

# ELEVEN

## From *Wissenschaftliche Philosophie* to Philosophy of Science

### 1. Introduction

Most current research on the origins of Logical Empiricism deals with developments before 1938. This is appropriate because that year marks the bitter end of scientific philosophy in Europe. With the *Anschluss* in March of 1938, Austria ceased to exist as a separate nation and Czechoslovakia was threatened. There was no place left in the German-speaking world for the scientific philosophers. By the end of 1938, almost everyone who was going to leave had done so.[1] Herbert Feigl, who had received his Ph.D. under Schlick in 1927, had already been at the University of Iowa since 1931. Carnap was appointed to the faculty at the University of Chicago in 1936, where he was soon joined, as a research assistant, by Carl Hempel. Reichenbach was on his way from Istanbul, where he had found refuge in 1933, to his new position at the University of California in Los Angeles. Philipp Frank was a guest of Percy Bridgman at Harvard.

It is not the mere fact of a historical break that has led those working on the history of Logical Empiricism to concentrate on the European phase of its development. Many of those engaged in this research are themselves products of the Logical Empiricist movement in its post–World War II forms. It is difficult for anyone to regard their own familiar professional development as history. The prewar period, by contrast, has until very recently remained largely unknown to contemporary philosophers of science, the subject of autobiographical remarks and disciplinary founder myths rather than of genuine historical scholarship.

The technical nature of scientific philosophy makes it difficult for anyone not trained in philosophy of science to investigate its development. Yet, in large part because their initial training has been as philosophers rather than historians, most of those now investigating the early history of Logical Empiricism tend to approach their subject as intellectual historians practicing what historians of science used to call "internal" history. The personal fortunes of the historical participants, as well as the larger social and cultural context of the time, remain in the background. Even these internal histories, however, contain indications of relevant background conditions. So there are fuller histories of this earlier period yet to be written. But one must begin somewhere, and the intellectual history is surely a good place to start.

If we now contemplate future studies focusing on the development of Logical Empiricism in North America, it is already clear that a rather different balance must be struck right from the start. The European origins of Logical Empiricism are not intellectually continuous with its later development in North America. But the cause of this discontinuity was clearly not primarily intellectual. It was the forcible dislocation of many of the major participants from the culture of German-speaking Europe during the interwar years to the English-speaking world of North America beginning around 1933. It is with this fact that any future history of Logical Empiricism in North America must begin.

The social facts are dramatic. In 1930, following publication of the Vienna Circle's manifesto "Wissenschaftliche Weltauffassung: Der Wiener Kreis" (Neurath 1929), these scientific philosophers represented just one among many modernist intellectual movements in the German-speaking world, and operated largely outside the German philosophical establishment. In 1960, just prior to publication of a volume entitled *Current Issues in the Philosophy of Science* (Feigl and Maxwell 1961), Logical Empiricism was clearly the dominant philosophy of science in North America.[2] So the overriding "external" question is this: How, between 1930 and 1960, did a dissident European movement advocating the replacement of much established German philosophy by *Wissenschaftliche Philosophie* transform itself into the dominant tradition for philosophy of science in North America?

As one seriously interested in the history of Logical Empiricism in North America, but not personally engaged in uncovering that history, I would like here to raise some general issues, pose some specific questions, and suggest some hypotheses that might be examined by future historians of this period. My motives are more than historical. Being able to see Logical Empiricism as the contingent historical development it surely must have been may allow future philosophers of science to put aside some old arguments and devote more of their energies to developing new approaches to understanding modern science. In addition, recovering the history may establish connections with earlier traditions containing forgotten resources useful to the contemporary enterprise.

## 2. Why Did the European Origins of Logical Empiricism Remain so Long in Relative Obscurity?

I begin with an issue already noted above. The European origins of Logical Empiricism remained in relative obscurity until well after it had lost its unchallenged position as *the* North American philosophy of science. Why was this so? The study of key texts has typically been a primary mechanism by which intellectual movements are transmitted. This was clearly not the case with Logical Empiricism. Carnap's *Aufbau* (1928), the major work to come out of the Vienna Circle, did not appear in English translation until 1967 (Carnap 1967), just three years before his death. Reichenbach's reputation in Germany rested primarily on his analyses of relativity theory as found in his *Relativitätstheorie und Erkenntnis Apriori* (1920), *Axiomatik der Relatativistischen Raum-Zeit-Lehre* (1924), and *Philosophie der Raum-Zeit-Lehre* (1928). It was this work that persuaded physicists such as

Einstein, Laue, and Planck to sponsor him for a position at the University of Berlin.[3] Yet none of these works appeared in English during Reichenbach's lifetime, with the latter appearing finally in 1958, five years after his death. Again, Schlick's major work *Allgemeine Erkenntnislehre* (1925) did not appear in English for fifty years (Schlick 1975). Wittgenstein's *Tractatus* (1922), by contrast, appeared in English translation already in 1922. Why was the case so different for the founding works of Logical Empiricism?[4]

My hypothesis is that the scientific philosophers, such as Carnap and Reichenbach, realized that their future, if they were to have a future, lay in North America. And they realized, quite rightly, that works like the *Aufbau* and *Relativitätstheorie*, which were written in the context of a cultural, scientific, and philosophical tradition that did not then exist in North America, would not be much appreciated in the North American context. So they put their efforts into other projects, ones better suited to their new intellectual and cultural environment. Carnap focused first on formal semantics (1942), and then on logical probability (1950), drawing on his work in semantics. Reichenbach devoted his Istanbul years to a general epistemological work, *Experience and Prediction* (1938), which he wrote in English to practice for his hoped-for emigration to the United States. Then he revised his book on probability for an English edition (1949a), also writing a logic text (1947) and technical treatises on quantum mechanics (1944), causality (1954), and the direction of time (1956)— the latter two published only after his death.

Even if I am mistaken about the motivations of leading Logical Empiricists during the 1930s and 1940s, the fact remains that the development of Logical Empiricism in North America proceeded in virtual ignorance of the major early writings that defined the European movement for scientific philosophy. Logical Empiricism in North America was to a considerable extent a new creation—built on the old foundations, to be sure, but styled for a new audience so that what appeared in public view in North America was something noticeably different from what had existed in Europe.

So what were the texts that first defined Logical Empiricism in North America? For the general philosopher and countless students, Ayer's *Language, Truth and Logic* (1936) provided an Anglicized version of the movement. Carnap's "Testability and Meaning" was originally published in the United States in 1936–37, and Reichenbach's *Experience and Prediction* in 1938. Then there were the early monographs in Neurath's *International Encyclopedia of Unified Science* (Neurath, Carnap, and Morris 1955), such as Carnap's *Foundations of Logic and Mathematics* (1939) and Ernst Nagel's *Principles of the Theory of Probability* (1939). And of course there were the major works on probability by the original leaders of the movement, Reichenbach's *The Theory of Probability* (1949a) and Carnap's *Logical Foundations of Probability* (1950). By the time the first volume of *Minnesota Studies in the Philosophy of Science* (Feigl and Scriven) appeared in 1956, Logical Empiricism had pretty well become the established philosophy of science in North America.

I suspect a large role was also played by the early *Readings:* Feigl and Sellars's *Readings in Philosophical Analysis* (1949) and Feigl and Brodbeck's *Readings in the Philosophy of Science* (1953). For the most part, the papers reprinted in these readers dated from the mid-1930s through the 1940s, after the migration to North America was well under way. Moreover, many of the articles by the scientific philosophers themselves were originally published in English, such as Carnap's "Testability and Meaning" (1936–37), Schlick's "Meaning and Verification" (1936), and Reichenbach's "The Philosophical Significance of the Theory of Relativity" (1949b). The paucity of writings by the original scientific philosophers before 1936 might be explained simply by the editors wanting to provide their readers with the most up-to-date materials. But how then do we account for Frege's "On Sense and Nominatum" (1892), Russell's "On Denoting" (1905), G. E. Moore's "Hume's Philosophy" (1922), C. I. Lewis's "The Pragmatic Conception of the *a Priori*" (1923), C. J. Ducasse's "Explanation, Mechanism, and Teleology" (1926), or P. W. Bridgman's "The Logic of Modern Physics" (1928)?[5]

It is a striking feature of these "readers" that they included papers by a number of Americans regarded as sympathetic to the movement: Lewis White Beck, May Brodbeck, R. M. Chisholm, Morris Cohen, C. J. Ducasse, Adolf Grünbaum, Sidney Hook, C. I. Lewis, Edward Madden, Paul Meehl, Ernest Nagel, W. V. O. Quine, Wilfrid Sellars, B. F. Skinner, K. W. Spence, W. T. Stace, and C. L.

Stevenson. Also well represented are British philosophers such as C. D. Broad, W. C. Kneale, G. E. Moore, and of course Bertrand Russell. No one can deny that the scientific philosophers were genuinely internationalist, outward-looking, and inclusive whenever possible. Moreover, they were explicit in advocating philosophy as a collective enterprise with many individuals contributing to its overall achievements. It is equally undeniable that this attitude promoted the institutional success of the movement in its new environment.

Another even more "external" explanation for why much of the spirit of the "Wissenschaftliche Weltauffassung" was left behind in the transition to North America is the simple but absolutely essential matter of institutional affiliation. The scientific philosophers had few institutional ties to philosophy in Europe. Reichenbach's position in Berlin, as noted, was in physics. When he left Europe, Carnap held a chair for Natural Philosophy in the Division of Natural Sciences at the German University in Prague. Feigl and Hempel never held any academic appointments in Europe. It was clear, however, that if they were to have academic positions in the United States, it would have to be as members of philosophy departments. To achieve this status they had to soften the rhetoric of *Wissenschaftliche Philosophie* and become Philosophers of Science.[6]

### 3. Hempel's Explications

For most academics, even most philosophers, the individual who best personified Logical Empiricism in North America was neither Carnap nor Reichenbach, but Carl Hempel. Hempel was originally Reichenbach's student in Berlin, but he also spent time in Vienna. He was caught with his dissertation not yet completed when Reichenbach was dismissed in 1933. He nevertheless completed his degree in Berlin the following year with Wolfgang Köhler, the Gestalt psychologist, serving as his official supervisor.[7] After spending some time in Belgium, he became Carnap's assistant at the University of Chicago in 1937 before moving to his first teaching position at Queens College in New York. Virtually all of his professional life was spent in the United States.

Hempel's early papers "Studies in the Logic of Confirmation"

(1945) and "Studies in the Logic of Explanation" (1948, with Paul Oppenheim) effectively defined what by 1960 were arguably the two most active areas of research in North American philosophy of science.[8] Neither topic had clear antecedents in the European writings of the scientific philosophers. Nor did the method employed, *explication*. That was a recent creation of Carnap's, first mentioned in the opening sections of "The Two Concepts of Probability" (1945a), of "On Inductive Logic" (1945b), and of *Meaning and Necessity* (1947, but written in 1942–44, 7–8), and discussed at length in the opening chapter ("On Explication") of *Logical Foundations of Probability* (1950).[9] Of course the idea of philosophy as the logical analysis of scientific concepts had long been central to scientific philosophy. But part of that idea, particularly for Schlick and Reichenbach, was that the concepts were connected with particular *scientific theories*, as the concepts of space and time are central to relativity theory. Explication, as practiced by Carnap and Hempel from the 1940s onward, had no such connection to any scientific theory. The concepts to be analyzed were general, methodological concepts supposedly common to all the sciences. Even in the case of confirmation, the fact that Carnap's final quantitative confirmation relation had the structure of mathematical probabilities was a *conclusion* of the analysis, not its starting point. For Hempel's purely qualitative confirmation relation, and for explanation, there is no associated scientific theory at all. The constraints on the analyses were provided by first-order logic and one's methodological intuitions expressed as initial conditions of adequacy for the resulting explicatum.

My questions here are: How does Carnap's 1940s notion of explication connect with his earlier views on the nature of philosophical analysis? What motivated him to develop this method in the way he did, when he did? These are questions of antecedents about which I know little. About the *consequences* of adopting the method of explication I have some definite hypotheses. One is that philosophical analysis as practiced by logicians and philosophers of science becomes comparable to philosophical analysis practiced by British "ordinary language" philosophers following the spirit of G. E. Moore and the later Wittgenstein.[10] One ends up comparing the analyses produced with prior intuitions through the construction of examples and counterexamples. The difference was that the result of explication is a new, more precise concept, a *replacement* for the original vague concept, more suitable for philosophical purposes. Wittgensteinian analysis, by contrast, supposedly leaves everything as it was, except that one should no longer be tempted to ask senseless questions. By the time of the Schilpp volume on Carnap's philosophy (1963), Carnap, ever tolerant, tried to see formal and informal analysis as complementary. All philosophy consists of conceptual analysis; only the method and tools employed may differ.[11]

Another consequence of regarding the explication of general methodological concepts as the primary task of the philosopher of science was an increasing separation between philosophy of science and the content of the sciences. People trained in philosophy, but with little knowledge of any science, could write article after article on "The Paradoxes of Confirmation" or "The Symmetry Between Explanation and Prediction." Carnap and Hempel cannot, of course, be blamed for the ensuing trivialization of much of the philosophy of science. No one can be expected to be that prescient. But when Kuhn came on the scene with a wealth of examples from the history of real science, the impact was amplified by comparison with the then existing discussions in the philosophy of science concerned with black ravens and the shadows of flagpoles.[12] As an indirect consequence of the challenge Kuhn posed for the philosophy of science, many philosophers of science in the 1970s attempted to reconnect the philosophy of science with real science. This effort was particularly noticeable in the philosophy of physics, in philosophical inquiries into the nature of probability and statistical inference, and a little later in the then newly reemerging philosophy of biology. To their eventual detriment, however, these efforts proceeded in virtual ignorance of debates over the proper relationship between science and philosophy that had taken place during the 1920s and 1930s.

### 4. Probability and Induction

Following publication of Reichenbach's *The Theory of Probability* (1949a) and Carnap's *Logical Foundations of Probability* (1950), probability and induction became major topics in the philosophy of science. The problem of induction had not played a significant role in the development of scientific philosophy before 1933. The scientific philosophers were of course much concerned with how experience and language connect with the world, and how the structure of experience could possibly reflect the structure of the world. But these questions were understood in a Kantian framework: How is objective scientific knowledge possible? They were not questions about the inductive warrant for claims about the real world. Even Carnap's *Aufbau,* we now realize, was primarily concerned with questions about the structure of objective knowledge, not about the empirical warrant for knowledge claims.[13] What accounts for the change in perspective?

My hypothesis is that a major factor in the change of perspective was the change in location and philosophical climate. Induction had been a major problem in the empiricist tradition since Hume, and, in spite of Kant, continued to be a major problem for Mill and then for Russell. It had ceased to be a problem in German philosophy after Kant. The Hegelian influence both on British philosophy and on the American Pragmatists had pretty much been overcome by Moore and Russell, and by Dewey, by the 1930s. Traditional British empiricism had reasserted itself. It was this tradition, I think, into which both Carnap and Reichenbach inserted themselves. Whether this was in any sense part of a conscious effort at assimilation I do not know. It was, nevertheless, a highly adaptive course of action.

Carnap recounts that he began to think more systematically "about the problems of probability and induction" in the spring of 1941 while on leave from the University of Chicago at Harvard University. He recalls lectures by Richard von Mises and by Feigl as being especially influential (Schilpp 1963, 36). It was in the spring of that year that he "began to reconsider the whole problem of probability" (Schilpp 1963, 72), a reconsideration that led him to reread Keynes's *Treatise on Probability* (1921). The result was first his paper "On Inductive Logic" (1945b) and then *Logical Foundations of Probability* (1950). I do not know of any further insights into his motivations in the published literature.

Reichenbach's case is more complex. The following story is told about Reichenbach. When the Nazis marched in and shut down the University of Berlin, Reichenbach is said to have exclaimed (a German equivalent of): "Now I understand the problem of induction!"[14] One thinks immediately of Russell's story of the man and the chicken in chapter 6 ("On Induction") of *The Problems of Philosophy* (1912). This story about Reichenbach may well not be true. But it is true that Reichenbach's published work on induction dates from his 1933 paper, "Die logischen Grundlagen des Wahrscheinlichkeitsbegriffs," which contains an outline of his famous "pragmatic" justification of induction.[15] It also seems to be true that Reichenbach was very proud of this argument and wanted to include it in a monograph on probability in the *International Encyclopedia of Unified Science*. But the editors, including Carnap and Neurath, had already engaged Ernest Nagel to do the monograph on probability (Nagel 1939) and wanted Reichenbach to do one on space, time, and relativity. He refused. As a result there was no contribution by Reichenbach on any subject.[16] I suspect that Reichenbach was consciously weighing what sort of publication would best further his prospects for a professorship in the United States.[17]

Probability is another matter. Reichenbach's dissertation, completed in 1915, was on the application of probability to the physical world. His *Wahrscheinlichkeitslehre* was published in the Netherlands in 1935. The problem of induction in its Humean form appears only in the final section, in which he outlines his pragmatic justification. This last section could well have been added after Reichenbach left Berlin.[18] In any case, this final section (Sec. 80), comprising ten pages, grew into a completely new chapter 11 ("Induction"), comprising fifty three pages, in the 1949 English edition. In the preface, dated May, 1948, Reichenbach refers to his proof of the justification of induction as originating "some fifteen years ago," which would put it in 1933. Reichenbach and Carnap engaged in both cooperation and competition all their professional lives. I do not know the extent to which either operated in the years immediately preceding publication of their books on probability. The controversy generated after publication in fact contributed positively to both their reputations, and to the vitality of Logical Empiricism, in North America.

## 5. Discovery versus Justification

At the end of the opening chapter ("A Role for History") of *The Structure of Scientific Revolutions* (1962), Kuhn notes that his remarks "may even seem to have violated the very influential contemporary distinction between 'the context of discovery' and 'the context of justification.'" These are about the only explicit references to Logical Empiricism that appear in his book. Referring to this and other unnamed distinctions, he goes on to say: "Rather than being elementary logical or methodological distinctions, which would thus be prior to the analysis of scientific knowledge, they now seem integral parts of a traditional set of substantial answers to the very questions upon which they have been deployed." I think Kuhn had it exactly right. The discovery-justification distinction was a substantial presupposition of the Logical Empiricist understanding of science as it developed in North America. He was wrong, however, in thinking it contemporary, and he underestimated the depth of commitment the distinction commanded.

Ironically, most of the critical literature on this distinction implicitly honors it by considering only its legitimacy and not inquiring into its origins. The source of the distinction "context of discovery"/"context of justification," formulated in just these words, is, as everyone recognizes, Hans Reichenbach's *Experience and Prediction,* published by the University of Chicago Press in 1938, almost certainly through the good offices of Charles Morris. As noted above, it was written in English during the years 1934–37 at the University of Istanbul where Reichenbach, along with fifty or so other former German professors (including Richard von Mises), found refuge in Mustafa Kemal's new Republic after being dismissed from his post in Berlin in early 1933. But versions of the distinction had been common in German philosophy for at least fifty years. The neo-Kantian Rudolph Hermann Lotze, for example, used a version in the 1870s to distinguish questions regarding the psychological genesis of spatial representations from questions regarding the validity of geometrical knowledge of such representations (Hatfield 1990, 163–64).[19]

Reichenbach introduces his version of the distinction in the very first section of *Experience and Prediction,* where he sets out the task of epistemology as he conceives it. Here he is primarily concerned to distinguish epistemology from psychology. This he does by associating the concerns of epistemology with those of logic, and then drawing on the long tradition, emphasized by Frege, of distinguishing logic from psychology. It will, he writes, "never be a permissible objection to an epistemological construction that actual thinking does not conform to it" (p. 6). This presentation proceeds in a manner suggesting that Reichenbach does not think his distinction needs much defense, and he does not seriously attempt to provide one.

Interestingly, the distinction reappears only once in *Experience and Prediction,* and then only briefly, near the end of the final chapter on probability and induction. Here he writes:

> What we wish to point out with our theory of induction is the logical relation of the new theory to the known facts. We do not insist that the discovery of the new theory is performed by a reflection of a kind similar to our expositions; we do not maintain anything about the question of how it is performed—what we maintain is nothing but a relation of a theory to facts, *independent of the man who found the theory.* (p. 382, emphasis added)

He then cites the example of Einstein and the general theory of relativity. The contrast between "the man who found the theory" and a logical "relation of theory to facts" exactly parallels that between the contexts of discovery and justification. My conjecture is that part of the significance of the distinction for Reichenbach at this time was its implicit denial that the character of a person proposing a scientific hypotheses has anything to do with the scientific validity of the hypothesis proposed. This applies, in particular, to being a Jew, from which it follows that his dismissal from his position in Berlin, as well as the persecution leading to Einstein's resignation, had been in principle unwarranted.[20]

My only supporting textual evidence for this conjecture appears in the final paragraph of a paper on the state of "Logistic Empiricism in Germany" which Reichenbach published in 1936 in *The Journal of Philosophy.* This paper is as much an advertisement for his own views, and a criticism of Carnap's views, as it is a survey of the movement he champions. There he writes: "Science, surely, is not limited to national or racial boundaries; we prefer to stand for this historical truth, in spite of all the pretensions of a certain modern nationalism" (Reichenbach 1936, 160). The "racial boundaries" include those between Jews and others, and the "certain modern nationalism" can only be National Socialism. In philosophical terms, Reichenbach seems to have made it a precondition for any scientific epistemology that it rule out the possibility of Jewish—or any culturally identifiable—science. But more than this, it seems that separating questions of the origins of ideas from questions of their validity was, for Reichenbach at that time, a matter as deeply personal as it was philosophical. And this sentiment was surely shared by everyone in the movement.

If one is going to insist on so strong a distinction between discovery and justification, one is obliged to produce a theory of justification to back it up. That Reichenbach did. His own theory of induction does satisfy the precondition that the justification of a hypothesis be independent of its origin. His rule of induction operates as a relationship between purely formal aspects of a fixed set of data and a single hypothesis—a relative frequency in a finite sequence of occurrences and a postulated limiting relative frequency, respectively. There is simply no place in such a formal relationship for any aspects of the wider context to enter into the calculation. The very different accounts of scientific inference developed by Carnap, and by Popper, also satisfy Reichenbach's requirement, and for similar reasons.

It is interesting to speculate on what would have been Reichenbach's reaction to N. R. Hanson's writings on the "logic of discovery" around 1960 (Hanson 1958, 1961). Hanson took Reichenbach's distinction to imply, for example, that Kepler was not thinking "logically" or "rationally" as he worked his way toward the hypotheses we now know as Kepler's Laws. In retrospect, this seems a gross misunderstanding of Reichenbach's intent. Reichenbach was concerned to distinguish, as was common among German philosophers since Kant, between psychology and logic. And he wanted to assert that the personal characteristics of scientists are irrelevant to the validity of their ideas. Claiming that there was no rationale in Kepler's attempts to formulate the laws of planetary motion seems to have been far from his intent.

There may well exist additional documentary evidence regarding Reichenbach's personal motivations for insisting on a distinction between discovery and justification around 1935. I doubt that he explicitly foresaw, and maybe he was never consciously aware of, the usefulness of such a distinction for a German immigrant in the United States in the subsequent years. For the distinction says: Don't think about the fact that I am a German immigrant, or speak with an accent; just consider the validity of my ideas. That had to be a very useful stance for anyone in his position.

## 6. What happened to Pragmatism?

The story of the success of Logical Empiricism in North America coincides with the story of the decline of American Pragmatism. In 1930, Pragmatism was the leading, though not the only, philosophical program in the United States.[21] By 1960 it was much diminished as an identifiable philosophical school and was no longer well represented in leading departments of philosophy.[22] The historical issue here is the extent to which these two phenomena were causally related. In particular, how much did the success of Logical Empiricism contribute to the decline of Pragmatism? Could these have been cotemporal but causally disconnected phenomena? My hypothesis is that they were somewhat connected in the way suggested, but that there were also independent factors at work.

It is a matter of historical record that members of both movements initially viewed each other as philosophical allies. Philosophers identified with Pragmatism, including Charles Morris, Ernest Nagel, and W. V. O. Quine, visited their European counterparts in the 1930s. These same people were soon thereafter instrumental in securing academic positions in America for their erstwhile hosts, including both Carnap and Reichenbach. Russell (1939) and Reichenbach (1939) both contributed to the first Schilpp volume, *The Philosophy of John Dewey*. And selections from the writings of Dewey and several other pragmatists were included in Feigl's *Readings*. Morris, in particular, sought to unify the two movements both through his writings (Morris 1937) and by becoming an editor and sponsor of Neurath's *International Encyclopedia of Unified Science*. So what happened in the following decades to change the philosophical climate so dramatically?

In spite of similarities, there were also deep differences between Pragmatism and scientific philosophy, including differences concerning the nature of philosophy itself. By 1929, when he turned seventy, Dewey was a philosophical naturalist, and to some extent even an evolutionary naturalist (Dewey 1910). There was, for Dewey, no special sort of philosophical knowledge, particularly none that could provide any sort of foundation or ultimate legitimization for the sciences. Rather, our understanding of evolutionary biology and psychology provided a basis for an understanding of scientific inquiry itself. What was special to philosophy, for Dewey, was the task of bringing to moral and political inquiries the conclusions and methods of the sciences. Moreover, he was far less concerned with the truth of scientific conclusions than with their usefulness for solving current societal problems.

In retrospect, one could describe the program of scientific philosophy as one of naturalizing that part of philosophy consisting of Kantian, or neo-Kantian, metaphysics. More specifically, in a scientific philosophy, an autonomous philosophical understanding of arithmetic and geometry, of space, time, and causality, is abandoned in favor of a *scientific* understanding of these concepts. But the rest of philosophy is not naturalized; it is transformed. Scientific philosophy becomes the logical analysis of the language, concepts, and theories of the sciences, an enterprise that, like modern mathematical logic itself, takes place in the philosophically autonomous realm of the analytic a priori. And this, for the Logical Empiricists, came to include scientific epistemology. That conception is clearest in the case of Carnap's inductive logic, but holds also for important aspects of Reichenbach's more pragmatic, decision oriented account of primary induction.[23]

Part of our question, then, is: How did a naturalistic pragmatism incorporating an empirical theory of inquiry get replaced by a philosophy that regarded induction as a formal relationship between evidence and hypothesis? In the 1950s, many philosophers of science would have given such a question very short shrift. Many would simply have asserted that Pragmatism was mistaken and Logical Empiricism correct. Induction just *is* a formal relationship between evidence and hypotheses. That is why Logical Empiricism took over. But such a response will not suffice in the 1990s. Ever since Quine advocated naturalizing epistemology (Quine 1969), philosophical sentiment has been moving back in Dewey's direction. So the question now is why philosophers then *believed* that Pragmatism was so obviously wrong and Logical Empiricism so obviously right.

Answering this question will require a close historical look at philosophical debates in the 1940s and 50s. What, for example, was the reaction of various philosophers to the opposition between the pragmatic conception of truth and the combination of Tarski's definition of "True in L" together with Carnap's (1949) distinction between truth and confirmation? Reichenbach (1939) accused Dewey, as well as Peirce, of failing to distinguish between the contexts of discovery and justification. Was this judgment widely shared and taken as a ground for dismissing Dewey's *Logic: The Theory of Inquiry* (1938)—published the same year as Reichenbach's own *Experience and Prediction* (1938)?

More "external" explanations are also possible. One possibility is that Pragmatism had played itself out. There may simply have been a lack of interesting new problems to engage the attention of new graduate students. Logical Empiricism, by contrast, came to North America with a large budget of problems and new logical techniques with which to attack them. So we may have here a case of a stagnant research program overwhelmed by a vigorous new program. Again, following a decade of economic depression and four years of war, the postwar period was one in which there was a strong desire in all areas of life to put the past behind and begin anew. Pragmatism was past; Logical Empiricism was new.[24] Moreover, Logical Empiricism identified itself with the new physics which, by making possible a nuclear bomb, had ended the war in the Pacific. It fit in well with the generally positive image of science following the war.[25]

Other less innocent cultural factors may also be relevant. Pragmatism, particularly as exemplified by Dewey, was a social philosophy. As such, it was bound up with various American social movements in the 1930s. Dewey himself visited Japan and China in 1919. He also visited the Soviet Union in 1928 and later published a series of articles which earned him labels like "Bolshevik" and "Red" in conservative segments of the American press. Then, in the mid-1930s, he went to Mexico as head of "The Commission of Inquiry into the Charges against Leon Trotsky." The verdict of "Not Guilty" led others to label him a "Trotskyite." [26] Later, in 1939, Dewey argued against American involvement in the war (Dewey 1939). Could it be that Pragmatism exhausted itself in the social and political battles of this period? [27]

More ominously, could Pragmatism to some extent have been a victim of anticommunism and McCarthyism in the decade following the war? Those are just the years when the balance between Pragmatism and Logical Empiricism seems to have shifted strongly toward Logical Empiricism.[28] If this is at all correct, there is considerable irony in the situation. In the European context, the scientific philosophers were socially every bit as radical as the American Pragmatists. Many had ties to European socialist and communist parties. That might initially have been part of the bond among members of the two movements. But these facts were left buried in the past. Once in America, the Logical Empiricist philosophers of science pretty much stuck to their Ps and Qs.[29] Their recent experience, after all, had surely convinced them of the destructive power of nationalistic political movements.[30]

## 7. The Next Generation

The primary means by which intellectual movements grow in the academic world is through the recruitment of graduate students who will constitute the next generation of professors. For whatever reason, following the war, the Pragmatists were not very successful in recruiting or placing new students. By contrast, even an impressionistic survey reveals the Logical Empiricists to have been extraordinarily successful.[31] Among Reichenbach's students at UCLA were Hilary Putnam, later long-time Professor of Philosophy at Harvard, and Wesley Salmon, who himself had many students both at Indiana University and at the University of Pittsburgh. Carnap had few students at Chicago, but among them was Richard Jeffrey, who eventually went to Princeton. While still at Queens College, Hempel taught both Adolf Grünbaum and Nicholas Rescher, both of whom had emigrated from Germany with their parents in the 1930s. Hempel later moved to Yale, where he became Grünbaum's dissertation director, and then to Princeton, where he had many more students. Grünbaum and Rescher both had many students at the University of Pittsburgh. Feigl, of course, had many students at Minnesota. Among American followers, Quine at Harvard and Nagel at Columbia both had many students. Among Nagel's students were Patrick Suppes, who himself trained many students at Stanford, Henry Kyburg at the University of Rochester, and Isaac Levi, who ended up back at Columbia.[32] The picture would be even richer if one considered less central figures such as Gustav Bergmann, Max Black, and Arthur Pap.

That a number of students of the early Logical Empiricists ended up in positions where they could also train graduate students is itself of considerable significance. When the postwar baby boom generation reached college age in the 1960s, universities expanded, and the new positions in philosophy thus created were filled by analytic philosophers and philosophers of science trained in the tradition of Logical Empiricism. The continued vitality of Logical Empiricism into the 1980s owed much to this further consequence of World War II.

In this regard, it is enlightening to contrast the fortunes of Logical Empiricism as an academic philosophy of science with the scientific philosophy of "The Vienna Circle in Exile" led by Philipp Frank at Harvard beginning in 1939 (Holton 1992, 1993, 1995).[33] Frank first formed a series of meetings including scholars from many fields, but mainly the sciences, under the heading "Inter-Scientific Discussion Group." In 1947 he created the Institute for the Unity of Science under the auspices of the American Academy of Arts and Sciences in Boston. Meetings and conferences were held, many attended by eminent scholars such as George Birkhoff, Wassily Leontief, Harlow Shapley, B. F. Skinner, and Norbert Wiener. The only possibility for training graduate students in the philosophy of science, however, was through Quine in the Department of Philosophy. So Frank's direct influence on subsequent developments in the philosophy of science was minimized. But his indirect influence was great. When Frank's Institute was dissolved in 1958, it was succeeded by what became the Boston Colloquium for Philosophy of Science, organized by Robert Cohen and Marx Wartofsky at Boston University.[34]

There is a common mythology, more common outside philosophy than within, that following publication of Kuhn's *Structure of Scientific Revolutions* there was a dramatic shift toward a more historical and less logical approach to the philosophy of science. The above analysis shows why this has to be a myth. By the 1960s, the majority of philosophers of science had been trained in the tradition of Logical Empiricism with its emphasis on logical analysis. One cannot quickly move from a logical to a historical mode of doing philosophy of science. A few philosophers of science shifted their focus to questions concerning conceptual change and the development of science, issues that had not been on the agenda of Logical Empiricism. Many more moved slowly away from the particular doctrines and projects of Logical Empiricism, often by moving closer to work in the sciences themselves. Thus, although in the 1990s there are very few philosophers of science who would identify themselves as Logical Empiricists, the majority are still pursing topics and employing means of analysis that are historically continuous with those of Logical Empiricism. That may be the best measure of the success of Logical Empiricism as a philosophy of science.[35]

## 8. Conclusion

In this look at the development of Logical Empiricism in North America, I have emphasized context over content. This is justifiable because many readers will already be familiar with the content and have thought little about the context. A satisfactory history of these developments will, of course, have to put the content into the context. But so as not to lose sight of the importance of the context, I would recommend one keep in mind the following related historical counterfactuals: Imagine that the Social Democrats rather than the National Socialists had come to power in Germany in 1933 (and thus that World War II never happened). What would have been the fate of *Wissenschaftliche Philosophie* in Germany, Austria, and throughout the world? What would have been the fate of American Pragmatism? And what would now be the complexion of the philosophy of science in North America?

# Conclusion

*Underdetermination, Relativism, and Perspectival Realism*

## 1. Introduction

By way of conclusion, I will briefly take up a well-known issue in science studies that has not been treated in any of the previous chapters. Seeing this issue from the perspective provided by my understanding of the practice of science provides a good way of reviewing the main features of that perspective.

The issue goes under the name "underdetermination," or, more informatively, "the underdetermination of theory by evidence." The supposed fact of underdetermination is widely taken to support various forms of social constructivism, and relativism more generally. The argument, often implicit, goes roughly like this. Scientific communities typically put forward particular theories as true, correct, accepted, the best available, etc. But theories are underdetermined by empirical evidence. So the evidence does not provide a sufficient basis for singling out the favored theory over other possible theories. There must, therefore, be something else which at least helps to explain why or how the favored theory was chosen. Since science is a social activity, the something else is most likely something social, such as social interests or social interactions. So some form of social constructivism is necessarily part of any fully adequate account of scientific practice.

A key premise in this argument is, of course, the underdetermination thesis. To evaluate the argument one must first get clearer as to the meaning of this thesis. In fact, the clearest expression of the thesis presupposes a standard, objectivist account of reference and truth, and a view of scientific theories as being simply sets of statements. The implications of underdetermination are not nearly so dire if one understands theories as providing perspectives within which one can construct models that fit the world more or less well.

## 2. Underdetermination and Objective Truth

The thesis of underdetermination can be clarified by considering an artificial universe consisting of a finite number of objects labeled **a, b, c, ..., n**. These objects may each exhibit a finite number of properties **F, G, H, ..., Z**. They may also participate in a number of relations, such as the binary relation **R(x,y)**. In short, this universe has a determinate set-theoretical structure which can be described by a finite number of statements. These might even include the "law" that all **F**'s are **G**'s. This whole set of statements constitutes the objectively true theory, **T,** of our universe. That there is such a theory is the view Hilary Putnam (1978) long ago dubbed *metaphysical realism*.

Now consider a proper subset of this universe described by the singular statements **Fa, Ga, Gb,** and **Hc**. Regard this set of statements as constituting empirical evidence, **E**. (See fig. C.1.) It is now a trivial truth of logic that, although **T** logically implies **E**, **E** does not imply **T**. So we may say that **T** is *logically,* or *deductively,* underdetermined by **E**. More generally, any evidence that fails to be exhaustive of the universe deductively underdetermines the true structure of that universe. Many casual references to the underdetermination of theories by evidence seem to involve only this trivial, although undeniable, form of *deductive* underdetermination.

Logic also provides the somewhat more interesting result that any non-exhaustive set of evidence fails logically to imply *any* description of the universe that goes beyond the description of that

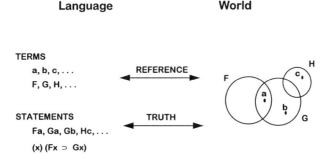

C.1 A standard, objectivist account of reference and truth.

evidence. So, not only does evidence logically underdetermine the *true* theory of the universe, it deductively underdetermines *any* theory of the universe as a whole. That is a somewhat more charitable reading of casual references to the underdetermination of theory by evidence.

A still more sophisticated version of logical underdetermination results if one introduces a distinction between properties that are *observable* and those that are not. Then the claim that theories are logically underdetermined by evidence need not refer to particular sets of data. Rather, theories are logically underdetermined by *all possible* evidence concerning to the observable properties of objects. But, as we have seen, having an observation-theory distinction is not essential to claims of deductive underdetermination. So one cannot disarm charges of underdetermination merely by discrediting the observation-theory distinction.

The thesis of deductive underdetermination well serves the interests of social constructivists and relativism in general. The only way to avoid logical underdetermination is to restrict one's theoretical claims to the evidence itself, or, more liberally, to what is in principle observable. But that presumes far too restrictive an understanding of scientific practice. Few, if any, scientific communities practice so austere an empiricism. So it looks like the relativist argument outlined above works for most of modern science. But it works so well only for *deductive* underdetermination.

Except for Karl Popper and his followers, few philosophers of science during the twentieth century have held that deduction alone exhausts the methods of science. Scientists, it is widely held, use *inductive* methods as well. It is also generally agreed that no inductive method can guarantee a determination of the true theory. At most, inductive methods yield judgments about the acceptability or probability of theories. Granting the existence of inductive methods suggests the still more sophisticated thesis of the *methodological,* or *inductive,* underdetermination of theory by evidence; that is, the evidence *plus* inductive methods together fail to yield a uniquely acceptable (or probable) theory of the universe.

Beginning around 1950, the major founders of Logical Empiricism in America, Carnap and Reichenbach, promoted programs for developing fundamental forms of induction that, if successful, would have refuted the thesis of inductive underdetermination. These, and related, programs continued to be developed with considerable vigor for several decades, and continue even today. Yet few would claim that success is imminent. The rival historical school in the philosophy of science, emphasizing the clash of competing research traditions, never even tried to develop methodological principles strong enough to defeat inductive underdetermination. Even more telling, no philosopher of science ever seems to have argued that any of the methods actually used in any of the sciences could rule out inductive underdetermination. It seems, therefore, that the resources of contemporary philosophy of science are, after all, insufficient to block the relativist argument from underdetermination.

## 3. Perspectival Realism

The above presentation of the underdetermination thesis and its implications relied on a standard, objectivist account of reference and truth. What makes the argument for relativism so compelling, and damaging, is the idea that there is a unique true theory of the world, but that our best methods fail to provide sufficient assurance that we have picked the true theory, or even that we are moving in the right direction, however that might be defined.

My view is that we should give up using the standard framework of objectivist reference and truth as a basis for developing an interpretation of the practice of modern science. Rather than thinking of science as producing sets of statements that are true or false in the standard objectivist fashion, we should think of it as a practice that produces models of the world that may fit the world more or less well in something like the way maps fit the world more or less well. In such a framework, it is sufficient that empirical evidence can sometimes help us decide that one type of model fits better than another type in some important respects. The fact that the process leading to the consideration and choice of the successful model included cultural values or presuppositions does not count against it fitting as well as we judge it does. It only means that there may be other models, or other types of models, that, if considered, might be judged to be even better fitting in the respects of interest.

The result is a kind of realism regarding the application of models to the real world, but it is a realism that is perspectival rather than objective or metaphysical. The sorts of general principles operative in some sciences provide a perspective within which particular models may be constructed. When, through observation or experimentation, these particular models are judged to be well-fitting, we are justifiably confident that the world itself exhibits a structure similar to that of our models. Realism need not require that we be in possession of a perfect model that exactly mirrors the structure of the world in all respects and to a perfect degree of accuracy. So the fact of underdetermination need not leave us in an unacceptable state of relativism.

# Notes

Introduction

1. This polarization is clearly exhibited in the major sustained work to date from the scientific side, Gross and Levitt's *Higher Superstition* (1994). See also Gross, Levitt, and Lewis (1996). Ross (1996) has edited what amounts to a reply from the side of the humanities and the social study of science. Here I provide only references applying specifically to this introduction. More extensive references are contained in the notes to individual chapters.

2. The distinction between *inter*disciplinary and *multi*disciplinary is defended in chapter 3.

3. This theme, for example, appears in a number of Herbert Feigl's essays (1981, 22, 40–41, 75, 367).

4. Historical studies of a related concept, *objectivity,* have recently appeared (Megill 1994).

5. My reference to the *intellectual* basis of the conflict suggests that there are other, *non*intellectual bases as well. And, indeed, this is certainly the case. There is *some* truth to the charge that some of those who reject scientific claims of genuine knowledge of the world have been motivated by a desire to undercut the authority of science. And this desire is often connected with a belief that scientific authority supports repressive social and political forces. On the other hand, scientists and their supporters desire the prestige and power that comes from a general belief that they possess uniquely rational methods that enable them to discover universal laws governing the fundamental workings of the natural world. Few today would follow Newton in maintaining that they are uncovering God's laws for nature, but the motivation is similar. My concern here, however, is with the *intellectual* bases of the conflict. If those bases can be eliminated, that may contribute to a lessening

of the overall conflict—even though intellectual agreement is hardly a guarantor of social or political agreement.

6. Not surprisingly, naturalism is also a concept with many meanings. My understanding of naturalism derives from debates within philosophy and the philosophy of science. For many sociologists, on the other hand, naturalism is the product of projecting social arrangements into nature, particularly biological nature. Philosophical naturalists may also be, but need not be, naturalists in this latter sense.

7. Some readers may recognize a similarity between these views and those expressed by Nancy Cartwright in *How the Laws of Physics Lie* (1983). Similarity there is. My way of putting things, however, is somewhat different. On my view, the kind of laws Cartwright is talking about cannot lie, but not because they tell the truth. Principles and definitions are not the sorts of thing that are even candidates for being either liars or truth-tellers about the real world.

1. Viewing Science

1. For Dewey's views on these issues, the best single work is probably his *Logic: The Theory of Inquiry* (1938), although his earlier *The Quest for Certainty* (1930) is far more readable. For a broader view of Dewey's work at that time, see Schilpp and Hahn (1939). Westbrook (1991) provides a very good overview of Dewey's whole life and work in the context of American culture.

2. For some recent contributions to this literature, and further references, see Giere and Richardson (1996). My afterword to this latter volume (Giere 1996c), here reprinted as chapter 11, includes an expanded version of the following remarks.

3. For examples, see the essays in Nickles (1980). Among the few works considering the history of the distinction are Hoyningen-Huene (1987) and Kusch (1995).

4. Another version appears in Karl Popper's *Logik der Forschung* (1935), but this version had little influence in North America before publication of the revised English edition in 1959 under the misleading title *The Logic of Scientific Discovery*.

5. Gerald Holton, for example, has recently written on the influence of Mach in America (1992) and "The Vienna Circle in Exile" (1993, 1995), which was organized by Philipp Frank at Harvard from the late 1930s into the 1950s. My question would be why Frank's group had so little influence on the strain of Logical Empiricism that got institutionalized as professional philosophy of science in succeeding years. For a more general cultural history of the period, including specific reference to Reichenbach, see Hollinger (1995).

6. Here I must note that philosophers of biology have been among those taking the history of science most seriously. The original leaders in the new philosophy of biology, such as David Hull and Michael Ruse, have also produced genuinely historical studies. This tradition continues among younger philosophers of biology.

7. Among prominent feminist empiricists are Helen Longino (1989) and Lynn Hankinson Nelson (1990).

8. David Hollinger (1983) has traced the evolution of Merton's paper from its

origins as a contribution to anti-fascist movements in the United States during the 1930s and 1940s to its emergence as a founding document for the development of sociology of science as a distinct field within sociology.

9. My category of *epistemological* constructivism includes the work of Harry Collins (1985) as well as the original works of the Edinburgh school. These include Shapin's (1975, 1979) study of phrenology in eighteenth century Edinburgh and Don Mackenzie's (1981) study of the development of statistics in Britain. The founding theoretical writings of the Edinburgh school include Barnes (1974, 1982), Bloor (1976), and the papers collected in Barnes and Shapin (1979), Barnes and Edge (1982), and Collins (1982). Shapin (1982) reviews many case studies from this perspective and provides a wealth of further references. Pickering (1992) provides a glimpse of later developments.

10. This was the view of Latour and Woolgar's *Laboratory Life* (1979), and it remains Woolgar's view (1988a). It is shared by Knorr-Cetina (1981) and many others. See, for example, the papers in Knorr-Cetina and Mulkay (1983). I comment further on Latour's later views in chapter 3. See also Pickering (1992).

11. Skepticism regarding the usefulness of the concept of a universal law of nature for understanding science may be found in works of Cartwright (1983, 1989) and van Fraassen (1980, 1989) as well as my own (Giere 1988, 1995). The latter essay is reprinted here as chapter 5.

12. For a more elaborate discussion of Putnam's proof, see Hale and Wright (1997).

2. Explaining Scientific Revolutions

1. For an elaboration of this line of thinking, together with quotations and historical references, see Rorty (1979, ch. 3). While admiring Rorty's historical analysis, I disagree with his rejection of representation as essential to knowing and, consequently, with his particular brand of pragmatism.

2. One might point to "creation science" and other forms of religious fundamentalism as showing that a philosophical justification of science is still much needed. In truth, the threat from religious fundamentalism is not so much intellectual as social and political. Nevertheless, for social and political reasons, it is important that the debate be joined, and prominent philosophers of science, such as Kitcher (1983) and Ruse (1982), have done so. What these philosophers have *not* attempted is to find a place completely outside of science from which to demonstrate that "creation science" is lousy science. That, I think, is impossible.

3. The following characterization of Logical Empiricism applies only to the version that developed in the United States following World War II. As recent scholarship shows, the European origins of Logical Empiricism were much more complex. For some indication of how complex they were, see Friedman (1991) and the essays in Giere and Richardson (1996). My own contribution to this latter volume, appearing here as chapter 11, treats the years after 1938, when the movement was first establishing itself in North America.

4. I owe this lovely term to Richard Burian.

5. The circumstance was that Kuhn had agreed to publish a monograph as

part of the *Encyclopedia of Unified Science*. These monographs were intended to be shorter even than Kuhn's book, which appeared as one of the last in a series that began as a principal organ of Logical Empiricism. Ironically, Kuhn's work turned out to be a major contributor to the demise of Logical Empiricism. By far the most extensive examination of Kuhn's work, although from a European perspective, is that of Hoyningen-Huene (1993).

6. Kuhn seems to have been influenced in Wittgensteinian ways of thinking by the philosophers Stanley Cavell and Paul Feyerabend, who were his colleagues at Berkeley in the late 1950s.

7. In this I follow Gerald Doppelt (1978), who, significantly for what follows, has also written on political philosophy. It must be admitted, however, that Kuhn himself was always fascinated by differences in linguistic practices among different research traditions. This comes out in his accounts of his earliest encounters with the history of science (Kuhn 1977). As someone trained in contemporary physics, he found he could not really understand the writings of much earlier scientists. But this evidence for linguistic incommensurability might be an artifact of the perspective of a scientist looking back at the written record of scientific changes that took place a hundred or more years ago. It does not follow that the *participants* in those changes experienced linguistic incommensurability. For another nonlinguistic account of incommensurability, see "The Anthropology of Incommensurability" (Biagioli 1993, ch. 4).

8. Among the more prominent members of the "historical school" in the philosophy of science are Lakatos (1978), Laudan (1977, 1984), Shapere (1984), and Toulmin (1961, 1972).

9. A noteworthy exception is Henry Frankel (1982, 1987).

10. For references, see chapter 1, note 9.

11. For references, see chapter 1, note 10.

12. A number of philosophers of science have been exploring accounts of science that, in one form or another, seem to embrace this combination. These include Boyd (1981), Churchland (1989), myself (Giere 1988), Hooker (1987, 1995), and Ruse (1986).

13. The political use of the word "revolution" in its modern sense seems not to have gained widespread currency until the later part of the seventeenth century (Cohen 1985, ch. 4). The word was at that time also applied to "the scientific revolution." Before that it had long been used to refer to any cyclical phenomenon, particularly astronomical phenomena, but also political phenomena. In this older sense of the term, "the structure of scientific revolutions" could refer to Kuhn's whole stage theory, in which the stages repeat, and not merely to the "revolutionary" stage. From this perspective, the title of the book may be more descriptive of the whole account than even its author intended.

14. For a comprehensive and critical review of the recent literature on evolutionary epistemology in general, see Bradie (1986). In addition to a wide-ranging set of papers, Callebaut and Pinxten (1987) provide an extensive bibliography, as do Hahlweg and Hooker (1989).

15. Left out of this survey is David Hull's (1988) heroic attempt to provide an evolutionary account of *both* the conceptual and the social evolution of science. Hull argues that organic evolution, the conceptual evolution of science, and its social evolution all exhibit the same generalized model of evolutionary change. This project generates the difficult problem of identifying isomorphic elements in all three realms and, by itself, fails to answer the causal question of what drives the evolutionary process. Hull's book is, nevertheless, necessary reading for anyone pursuing an evolutionary account of science.

16. For a brief historical survey of the literature on group selection see Sober (1984, 215–26).

17. The classic reference on punctuated equilibria is Gould and Eldredge (1977).

18. More congenial is the work of Herbert Simon and his associates on scientific discovery (Langley et al. 1987; Kulkarni and Simon 1988). Also promising is work based on systems with parallel distributed processing (Churchland 1989). For a survey of work on a cognitive approach to understanding scientific revolutions, see Giere (1991) and the papers in Giere (1992).

19. Such things were, of course, banned from behaviorist psychology as unscientifically "mental." The advent of the computer made that stance impossible.

20. This view of theories is developed in Giere (1988), chapters 3–5.

21. Standard references to the literature on human judgment include Kahneman, Slovic, and Tversky (1982) and Nisbett and Ross (1980).

22. The phrase is borrowed from Alan Franklin (1986, 1990). Others who have emphasized the role of experimentation in science include the philosophers of science Nancy Cartwright (1983), Ian Hacking (1983), and David Gooding (1990) and the historian of science Peter Galison (1987, 1997). See also Gooding, Pinch, and Schaffer (1989).

23. That discrimination is a major goal of experimentation is emphasized by Mayo (1996).

## 3. Science and Technology Studies

1. For illustrations of the application of models from Science Studies to Technology Studies see R. Laudan (1984) and Bijker, Hughes, and Pinch (1987). Justifications for the basic strategy are to be found in the introductions to both of these volumes. The introductory essay by Pinch and Bijker carries the revealing subtitle "How the Sociology of Science and the Sociology of Technology Might Benefit Each Other." Woolgar (1991) is critical of the whole strategy.

2. The connection between the rationality and the autonomy of science remains a central theme even in Dudley Shapere's (1984) historically based, postpositivist philosophy of science.

3. For the social uses of Newtonianism in England see Jacob (1976). The French Enlightenment is chronicled by Gay (1969). For a more recent overview, see Toulmin (1990).

4. See chapter 1, note 8, and associated text.

5. For representative essays supporting this view of technology, see Bijker, Hughes, and Pinch (1987) and MacKenzie and Wajcman (1985).

6. See chapter 1, notes 9 and 10.

7. The fact that most scientific and technological representations are eventually shown to be mistaken in detail does not refute this claim. It only shows that understanding scientific and technological representations in terms of simple truth or falsity is inadequate.

8. I can speak authoritatively only for myself (Giere 1988), but other philosophers of science who might also accept the label include Boyd (1981), Churchland (1989), Hooker (1987, 1995), and Kitcher (1993). I would also include Donald Campbell.

9. Here I would cite only the now classical works of Barnes (1974), Bloor (1976), and MacKenzie (1981).

10. This thesis is argued in the later chapters of Giere (1988). Interestingly, Woolgar (1991) agrees with my analysis of interest theories except in that he draws negative conclusions while I draw positive ones.

11. See (Latour 1993), where even the distinction between nature and society is ultimately rejected.

12. Hughes (1983, 1989). Note that there is no widely recognized counterpart to the systems approach in Science Studies more generally. Hooker (1987, 1995) probably comes the closest.

13. Pickering (1995, 215) characterizes this approach disparagingly as "eclectic multidisciplinarity" and proposes cultural studies as an "antidisciplinary new synthesis."

## 4. Naturalism and Realism

1. For recent criticisms of naturalism as either self-defeating or vacuous, see Plantinga (1993, chs. 11 and 12) and van Fraassen (1993).

2. The term "methodological naturalism" has been used by Plantinga (1995) and others to describe a possible strategy for scientists who are also theists. Plantinga himself rejects this strategy in favor of one that explicitly injects theistic components into one's scientific work. See also his (1991, 1996). I am recommending methodological naturalism as a strategy for scientists who are not theists and as a theoretical stance for philosophers of science.

3. Hilary Putnam (1982) provides a clear statement of this position.

4. Here I am thinking particularly of Barry Stroud (1996).

5. Oliver Sacks (1996) provides an engaging account of a Pacific island on which, due to an exceptional genetic history, the proportion of people exhibiting achromatopsia is one in twelve rather than the more normal 1/30,000. I thank Louise Anthony for this reference.

6. The idea of using maps as an analogue for scientific theories is hardly new, having been employed by recent thinkers as diverse as Polanyi (1958, 4) and Kuhn (1962, 108). It has been developed at greater length by Toulmin (1953, ch. 4) and Ziman (1978, ch. 4).

## 5. Science without Laws of Nature

1. Among recent philosophers, David Armstrong (1983) seems to me to come closest to the Kantian stance.

2. This answer is an obvious generalization of Reichenbach's (1938, ch. 1) famous distinction between the contexts of discovery and justification for scientific hypotheses. It may also be understood as seeing the excursion into history as committing a "genetic fallacy."

3. Armstrong (1983, 4), for example, writes: "If the discovery of the laws of nature is one of the three great traditional tasks of natural science, then the nature of a law of nature must be a central ontological concern for the philosophy of science." Similarly, John Earman (1986, 81) describes the concept of laws of nature as "a notion that is fundamental to the study not only of determinism but to the methodology and content of the sciences in general."

4. These features form a subset of the assumed characteristics of laws that van Fraassen (1989, 38) picks out as preeminent.

5. Here it is important to observe the medieval distinction between what is necessary for God's creations and what is necessary for the deity itself. Both Descartes and Newton were "voluntarists" in that they believed God could have chosen other laws for the world. Descartes notoriously even said that the laws of arithmetic and geometry could have been different if God had so willed.

6. Quoted by Westfall (1971, 397). I owe this reference to Brooke (1991, 139). Chapter 7 of Westfall's book, particularly the last five pages, makes a strong case for the role of Newton's conception of the deity in his willingness to abandon direct mechanical interaction in favor of apparent action at a distance.

7. As Milton notes (1998, n. 1), "There is no satisfactory general history of the idea of a law of nature." That our modern use of the concept of "laws of nature" is directly traceable back to Descartes and Newton, and flowed from their conceptions of the Deity, was argued both by Zilsel (1942) and by Needham (1951). Both claimed that the idea of God's laws for nature originated with the rise of powerful centralized governments in the early modern period. Thus Zilsel (1942, 258) argues unequivocally that "the concept of physical law was not known before the seventeenth century" and suggests, more tentatively, that "the doctrine of universal natural laws of divine origin is possible only in a state with rational statute law and fully developed central sovereignty" (1942, 279). Oakley (1961; 1984, ch. 3) objects that the idea existed long before in a theological tradition. Ruby (1986, 350) argues that, already in the thirteenth century, Roger Bacon used the notion in a way that "resembled that of modern science." Milton (1981, 182) argues convincingly for the "suggestion that scientists discovered laws of nature because they were already thinking of nature in lawlike terms, and not the other way round . . ." While agreeing that the concept was used prior to the seventeenth century, Milton (1998, 699) concludes that "it was Descartes who more than anyone else created the modern idea of a law of nature, by conceiving the laws of nature specifically as laws of motion, and by making these laws the ultimate explanatory principles of his physics." Steinle (1995), surveying the notion of a law of nature in the works of Galileo,

Descartes, Bacon, Boyle, and Newton, reaches a similar conclusion, though insisting on significant differences within this group. Keller (1985, 131–35) and Harré (1993, 9–10) also note the seventeenth century religious origins of the concept of a law of nature.

8. I owe consideration of the importance of mathematics in this history to conversations with Rose-Mary Sargent. She also pointed out that Boyle's cautions regarding the use of the notion arose partly from his conviction that mathematical relationships abstracted too much from the complexities of nature. See her (1995).

9. The influence of Newton's conception of science on later British thought hardly needs documenting. For its influence on French Enlightenment thought, see Gay (1969, Book 3, ch. 3).

10. For an appreciation of the intensity of these debates see Brooke (1991, ch. 8) and Desmond and Moore (1991).

11. One might inquire into what forces sustained the interpretive role of the concept of laws of nature during and after its secularization in the nineteenth and twentieth centuries. One factor, I would guess, is that it well served the increasing professional interests of scientists, particularly physicists, to present themselves to a broader public as having the powers to discover universal laws of nature.

12. This thesis was argued thirty years ago by Michael Scriven (1961) and more recently by others including Nancy Cartwright (1983, 1989) and myself (1988). Even Armstrong (1983, 6–7) and Earman (1986, 80–81) admit the strict falsity of the traditional examples of laws.

13. For a more extended discussion of the strict falsity of the law of the pendulum, see Giere (1988, 76–78). That classical mechanics has been superseded by relativity theory and quantum mechanics does not materially change the argument. Cartwright (1983, 1989) provides examples from quantum theory. Similar examples could surely be developed for relativity theory as well.

14. For an elaboration of this view, see Swartz (1985).

15. For a more extended development of this objection, see my paper "Laws, Theories, and Generalizations" in Grünbaum and Salmon (1988, 37–46).

16. Cartwright (1983) holds the superficially similar view that lower-level laws, such as Snell's Law, are to be understood as *ceteris paribus* laws of the form: "Everything else being equal, . . . , etc." But she does not claim that such laws are true, only that they are explanatory in a way not compatible with a covering law model of explanation. I would prefer a more radical interpretation that does away with law talk even though this departs from the way scientists themselves often present their science. I think this can provide us (philosophers and historians) with a better understanding of what they (scientists) are doing.

17. Van Fraassen (1980, 1989), for example, defines scientific realism as implying that one of a family of models is exactly isomorphic with the system it is intended to represent. I have objected (Giere 1985; 1988, ch. 4) that this is too strong a requirement for a reasonable realism. See chapter 9 of the present volume for a version of this argument.

18. Recall also that the title of Descartes's main work in natural philosophy

translates as *The Principles of Philosophy*. Edith Sylla tells me, however, that in the seventeenth century "principle" was typically used to refer to a general *truth*.

19. It is worth noting that in the twentieth century the expression "the principle of relativity" has had considerable currency, as in the title of the well-known collection of fundamental papers by Einstein and others (Lorentz et al. 1923). Einstein himself (1934) distinguished between what he called "constructive" theories and "principle" theories. The special theory of relativity, he claimed, was of this latter type. One of its principles is the "principle of the constancy of light in vacuo" (1934, 56). Einstein describes the advantages of principle theories over constructive theories as being "logical perfection and security of the foundations" (1934, 54). Tom Ryckman, in a personal communication, suggests that Einstein's distinction might be close to my own. But it might also be the case that Einstein thought of his "principles" as expressing deep general truths about the world, and, like Newton, drew on religious, though not theological, inspiration.

20. These two positions are represented by van Fraassen (1980, 1989) and Cartwright (1983, 1989) respectively.

21. Van Fraassen (1980, 1989) provides the archetype for "the empiricist" of this objection.

## 6. The Cognitive Structure of Scientific Theories

1. The roots of the model-based view go back to the work of J. C. C. McKinsey, Evert Beth, and John von Neumann in the 1930s, 40s, and 50s. It came to prominence in the philosophy of science in the 1960s, 70s, and 80s through their followers, Patrick Suppes (1979), Bas van Fraassen (1970b), and Frederick Suppe (1973, 1989), respectively. It is now prominent in Germany largely due to Suppes's student, Joseph Sneed (1971), and one of Carnap's followers, the late Wolfgang Stegmüller (1976, 1979), who was also influenced by Sneed. For a survey of recent European developments, see Baltzer and Moulines (1996). Suppe (1989) provides a good bibliography and a useful participant's perspective on these developments. In the original version of this chapter, I used the term "model-theoretic view" for this newer view of theories. I now prefer the term "model-based view," which is more neutral regarding the role of formal model theory in the application of the view. Harré (1987, 1993; Aronson, Harré and Way 1995) shares my skepticism regarding the usefulness of formal model theory for understanding scientific, as opposed to purely mathematical, theories. Clearly there are several variations on this theme, of which my own is only one.

2. For a sample, see Giere (1988), Stegmüller (1976, 1979), Suppe (1989), and van Fraassen (1980, 1989).

3. In *Explaining Science* (Giere 1988, ch. 3) I argued that theories are typically presented in scientific textbooks as families of models. I also speculated that the form of textbook presentations might have some basis in the general way humans perform tasks of cognitive organization, and I tried to make this speculation plausible by pointing to research on chess masters and problem solving by trained physicists. Nevertheless, my approach there suggests the existence of more substantial

connections between a model-based understanding of scientific theories and the cognitive sciences. This chapter is part of a project to explore such connections.

4. Smith (1990, 51) explicitly notes this "gap between the psychological and philosophical work on [categorization]."

5. There is also some evidence that central instances are learned earlier than peripheral instances (Rosch 1973b), further supporting the psychological reality of a gradation from central to peripheral cases.

6. In terms of biological classification, Rosch's basic level typically corresponds to the level of genus or family. One should not expect, however, that technical biological categories would neatly overlap the natural categories of everyday, folk classification.

7. Similarity theories of categorization are discussed in considerable detail by Smith and Medin (1981). For a recent review emphasizing more complex measures of similarity, see Smith (1990).

8. These lines of criticism are developed by Murphy and Medin (1985) and Medin (1989).

9. For recent work emphasizing the role of "theories" in categorization see Keil (1989), Murphy and Medin (1985), and Medin (1989).

10. Here one could identify models as either prototypes or exemplars. The usage of these terms is not that well delimited. I would prefer to call them "prototypes," saving the term "exemplar" for those special prototypes that play the particular historical and pedagogical roles that Kuhn (1962, 1974) assigns to his "exemplars."

11. The map is partial not just because it must fit on one page, but because, like all maps, it is necessarily partial. There could be no such thing as a "complete" presentation of all the models of classical mechanics.

12. Even the apparent exception, "Conservation of Energy," is misleading because these problems were really momentum problems, but the description contained a legitimate reference to energy, and the novice was misled.

13. I know of only one fairly extended discussion in Rosch's writings treating the effects of expertise on categorization (Rosch et al. 1976, 430–32). It suggests that expertise "lowers" the basic level. Her paradigm of an "expert" would be a judge for dog shows. For such an expert, "Collie" and "Setter" would be basic level categories; "Dog" would be superordinate. Such people do not see dogs as just dogs; they see them as collies, setters, etc. My suggestion does not necessarily imply that the basic level moves "up," only that experts in classical mechanics learn easily to categorize real systems as representable in terms of higher-level models, e.g., conservative models.

14. This is, of course, a controversial assumption going back to the work of Piaget. For a recent treatment based on current work in developmental cognitive psychology, see Carey (1985).

15. A possible exception is the Sneed-Stegmüller-Moulines tradition. In this tradition, it is a "constraint" on applications of models that the same objects be assigned the same parameters in different applications. The Earth, for example,

must be assigned the same mass both in models of the Moon-Earth system and in models of the Earth-Sun system. For a brief overview of this position, see Gähde (1989).

## 7. Visual Models and Scientific Judgment

1. The original version of this chapter, under the title "The Visual Presentation of Theory and Data: A Cognitive View," was presented at a meeting of the Society for Social Studies of Science in November 1987, and then for the Committee on History and Philosophy of Science at Johns Hopkins in December. In February 1989, a later version, under its present title, was presented at the Science Studies Units of the Universities of Bath and Edinburgh, and for the Department of Logic, Methodology, and Philosophy of Science at the University of London. A still later version, "Visual Models in Science: Lessons from the Revolution in Geology," was presented for the Centenary Conference on the History of Science at the University of Oklahoma in September 1990. The current version, once again extensively rewritten, was first published as Giere (1996a). I am especially grateful to Professor Brian Baigrie for providing the opportunity for me first to put these ideas into print.

2. Here it is worth noting that Kuhn's book contains not a single illustration.

3. For good examples of the initial philosophical reaction to Kuhn's work see Shapere (1964) and Scheffler (1967). Kuhn's implicit reply is scattered throughout the essays reprinted in *The Essential Tension* (1977).

4. For references, see chapter 1, notes 9 and 10.

5. For these developments within the sociology of science, see Lynch and Woolgar (1990), particularly the editors' introduction and the essays by Latour and Tibbetts. I myself (Giere 1994b) have reviewed these essays in some detail.

6. For references, see chapter 6, note 1.

7. One way of elaborating these ideas is explored in greater detail in chapter 6.

8. For a defense of the view that crucial experiments are typically after the fact reconstructions, see Brannigan (1981).

9. I have criticized probabilistic accounts of human judgment in chapter 6 of *Explaining Science* (1988) and Thagard's coherentist approach in Giere (1989c).

10. For a justification of the strategy of focusing on individuals, see Giere (1989a).

11. I have argued the virtues of an instrumental naturalistic rather than a categorical normative account of human judgment in *Explaining Science* (1988), especially chapters 1 and 6, and further defended this approach in Giere (1989b).

12. Sequential testing, however, introduces the possibility that which model ends up being chosen is a function of the particular order in which alternatives were considered.

13. In this account I have omitted (i) justification for the lack of prior probabilities in the model, (ii) explicit reference to the overall utilities of the decision maker, and (iii) the need for a supplemental decision strategy, such as satisficing, which justifies the obvious decision strategy in terms of satisficing relative to vari-

ous expected utilities. These aspects of crucial decisions are discussed in Giere (1983), in chapter 6 of *Explaining Science* (1988), and, in a more specialized context, in chapter 9 of the present volume.

14. Twentieth-century geology, particularly the 1960s "revolution" in geology, is fast becoming a standard test case for science studies. Among recent books in which it features in whole or in part are those by Le Grand (1988), Stewart (1990), Thagard (1991), and myself (Giere 1988, ch. 8). The articles are already too numerous to list here. One article (Le Grand 1990), however, deserves mention, as it explicitly uses images from the 1960s revolution in geology to argue the case for the importance of visual imagery in science.

15. Marvin (1973, 43), for example, reproduces an 1858 engraving by Antonio Snyder showing a much too good fit. Speculation regarding the fit of the two hemispheres seems to have followed shortly upon the production of maps of the new world comparable in detail to those of Europe and Africa. This fact fits Latour's (1986) thesis about the importance of "centers of calculation" for scientific progress. Direct comparisons of the coastlines become possible for a single observer only after many measurements have been brought together in one place and rendered graphically on a single piece of paper for easy viewing.

16. I suspect that Wegener may have been the first to emphasize such geological congruencies, but I am not myself sufficiently familiar with the historical sources to vouch for this suspicion.

17. Herbert Simon (1978) distinguishes "informational" from "computational" equivalence for representations. Informational equivalence is a generalization of logical equivalence. Computational equivalence means that the same information can be extracted from the representation with the same computational resources. In terms of this distinction, a linear, digital encoding of an image, as for television transmission, may be informationally equivalent to the reconstructed image on a screen. But these two representations are not computationally equivalent. In particular, an ordinary human would find it physically very difficult, if not physically impossible, to extract particular spatial information from the digital representation. But simply by looking at the pictorial representation, anyone could easily determine, for example, that the cat is on the mat. I would prefer a slightly different terminology, saying, rather, that the two representations are "logically" equivalent but not "cognitively" equivalent. The fundamental idea, however, is the same.

18. This image is taken from the first (1944) edition of Holmes's textbook (Holmes 1944, 506). A very similar diagram appeared in an article published in 1930 (Holmes 1930).

19. It is an interesting question why a dynamic version of Hess's model incorporating geomagnetic reversals does not appear in the literature until after 1966. My suspicion is that publishing conventions in professional journals like *Nature* at that time favored diagrams and graphs that presented *data*, as opposed to those that merely pictured speculative models.

## 8. Philosophy of Science Naturalized

1. Among prominent evolutionary or naturalistic epistemologists, I count Richard Boyd (1981), Donald Campbell (1974), Clifford Hooker (1987, 1995), and Abner Shimony (1971, 1981, 1993). Toulmin's (1972) view is evolutionary but perhaps not naturalistic. Popper (1972) appropriates the title "evolutionary" without adequate justification. Friedman (1979) reaches conclusions that are naturalistic but not evolutionary. Kitcher's (1993) scientific epistemology is likewise naturalistic but not evolutionary. Arthur Fine's (1984, 1986) "natural ontological attitude" encompasses a natural epistemological attitude as well. Outside the philosophy of science, advocates of naturalistic or evolutionary epistemologies are too numerous even to begin mentioning.

2. I first formulated this argument in Giere (1973) as an expression of what I then took to be a majority view among philosophers of science. I did not intend to argue that it was impossible to establish a connection between the philosophy of science and the historical practice of science; only that the authors under review had failed adequately to address the most serious difficulties. Indeed, my own solution at the time (Giere 1975) was basically naturalistic, though not evolutionary.

3. The importance of broadly empirical considerations in the Logical Empiricists' rejection of Russellian foundationism has been emphasized by Hempel (1983). In this paper, Hempel distinguishes "normative" from "descriptive-naturalistic" methodologies, and argues for a mixed approach.

4. For a summary of the relevant literature, and many references, see Giere (1979; 1988, ch. 6). Advocates of Bayesian inference seem to assume that their reconstruction of scientific inference, while not strictly reducible to deductive logic, nevertheless somehow carries the normative force associated with deductive logic. I do not understand the basis for this assumption. I am not even convinced that deductive logic possesses all the normative powers commonly ascribed to it.

5. The following discussion applies only to the Laudan of *Progress and Its Problems* (1977). He no longer subscribes to this metamethodology. Lakatos is somewhat ambiguous as to whether his metamethodology is just his ordinary methodology applied at the metalevel or a different methodology altogether.

6. What follows owes at least part of its inspiration to Paul Churchland's (1979) notion of an "epistemic engine." See also P. S. Churchland and P. M. Churchland (1983); P. S. Churchland (1986); P. M. Churchland (1989).

7. For some neurobiological findings relevant to the mechanisms underlying spatial coordination among mammals, see O'Keefe and Nadel (1978), and Pellionisz and Llinas (1982).

8. Here I am thinking particularly of Donald Campbell (1974) and his followers.

9. From informal comments at a conference in May 1984, I infer that Putnam himself might agree that his view is a variety of emergentism.

10. The label "constructive realism" was originally intended as a direct contrast to van Fraassen's (1980) "constructive empiricism." See Giere (1984) and

chapter 9 of this volume. My view of theories is a liberal version of his "semantic" conception of theories, and similar to Frederick Suppe's (1973, 1989) conception. Van Fraassen's distinction between "observable" and "theoretical" seems to me a philosophical imposition. It is very difficult to interpret the actual practice of scientists as honoring such a distinction. I find Nancy Cartwright's (1983) anti-realism much more congenial, perhaps even compatible with a constructive realism. There are some general similarities between my view and the "structuralist" approach of Sneed (1971) or Stegmüller (1979). This school, however, seems primarily interested in reconstruction and philosophical vindication, and less concerned with description.

11. Satisficing has been developed primarily by Herbert Simon. See his (1972) for further details and references.

12. For an example of the ambitious strategy, see L. Laudan (1984).

## 9. Constructive Realism

1. The quote is from van Fraassen (1980, 3). It will be obvious to readers that my debt to van Fraassen is enormous. I have benefited not only from his writings, but from several decades of discussion and correspondence that was unfailingly gracious and patient. In genesis, at least, without constructive empiricism there would be no constructive realism. The general idea of conjoining constructivism with realism into a constructive realism was later developed independently, and with different motivation, by Wallner (1990).

2. Suppes's more recent writings (1979) exhibit a more liberal view of the proper language for foundational studies. This liberal view now separates Suppes, Suppe (1973, 1989), and van Fraassen from the "structuralist" school of Sneed (1971) and Stegmüller (1976, 1979).

3. The writings of Putnam (1975a) and Boyd (1973, 1981), for example, are full of talk about approximation. But their sentential framework makes it difficult for them to deploy the notion effectively. This makes van Fraassen's criticisms seem more powerful than they are.

4. Van Fraassen mentions the studies of McKinsey, Sugar, and Suppes (1953) and Simon (1954).

5. He writes (1980, 57), "I regard what is observable as a theory-independent question."

6. The threatened vicious circle is that we must use psychology and physiology to tell us what are the observable substructures of our models of psychological and physiological systems.

7. The importance of detectability rather than observability has been emphasized by Churchland (1979) and Shapere (1982).

8. "The locus of possibility is the model, not a reality behind the phenomena" (1980, 202).

9. Constructive realism is thus a model-theoretic analogue of a view advocated by Grover Maxwell (1962).

10. Van Fraassen himself gives a sharp formulation of the empiricist-realist debate in terms of safety and strength in a later paper (1981).

11. The methodologies of Lakatos (1970) and Laudan (1977) may be understood as providing an empirical account of scientific activities such as the pursuit or acceptance of a theoretical program.

12. For an authoritative introduction to uses of satisficing, see Simon (1957).

13. For references and further discussion of a classical decision-theoretic approach to the acceptance of theories, see Giere (1983).

14. My understanding of the history of this case is derived from Watson (1968) and Olby (1974).

15. This holds even for the *modal* version of the realistic hypothesis.

16. My point here is similar, though in a different framework, to that argued by Boyd (1973, 1981).

17. I have adapted the formulation "security in strength" from William Harper.

18. The classical decision-theoretic approach thus provides an "inductivist" reconstruction of established Popperian dogma. Mayo (1996) develops a full-blown theory of inference based on the notion of a severe test.

19. The reader may wish to consult van Fraassen's own reply (1985, 289–94) to the above argument.

20. The phrase "theory of science" appears several times in *The Scientific Image*, e.g., on page 196. I am not sure that van Fraassen intended it to be taken as seriously as I am now taking it. In a later paper (1981), however, he states that the goal of the philosophy of science is to "make sense of science." Moreover, he explicitly employs his own empiricist conception of what "making sense" requires. So my description of his enterprise fits his later characterization even if this does not appear explicitly in *The Scientific Image*.

21. The requirement of reflexivity for the sociology of science is discussed, for example, by Bloor (1976).

22. By describing what he is doing as giving a more "accurate" description of scientific activities, van Fraassen disguises, even from himself, the degree to which he is *reinterpreting* these activities.

## 10. The Feminism Question in the Philosophy of Science

1. For an overview and references on this topic, see (Longino 1990, 106–11).

2. Gilligan's conclusions, of course, have been criticized right from the start (Walker 1984).

3. I say this is only a *possible* example because, although widely cited, for example, by McGrayne (1993), Keller's interpretation is open to serious criticism. The best I have seen is a draft manuscript by Nathaniel C. Comfort entitled "Bringing Up Barbara: The Maturation of Controlling Elements." Comfort distinguishes the phenomenon of transposition from McClintock's interpretation of that phenomenon. The phenomenon, he argues, using contemporaneous papers and letters, was widely accepted at the time, but the interpretation was, and continues to be, rejected as unclear or mistaken. Whatever the ultimate judgment on this case, Comfort's argument is at least pitched at an appropriate level of scientific and historical detail. On this much even Keller would surely agree.

4. I have developed these and related themes at greater length in Giere (1996c), reproduced as chapter 11 of this volume.

5. See chapter 1, note 3.

6. For further references and elaboration on model-based accounts of scientific theories, see chapter 6, note 1 and Giere (1988), chapter 3.

7. I find inspiration for both this terminology and the concept in some works of Donna Haraway, particularly her paper "Situated Knowledges: The Science Question in Feminism and the Privilege of Partial Perspective," reprinted in (Haraway 1991). For somewhat more elaboration, see chapter 4 of the present volume.

8. Among feminist *epistemologists*, however, Naomi Scheman (1993) may be counted as a feminist realist, even a perspectival realist.

## 11. From *Wissenschaftliche Philosophie* to Philosophy of Science

1. Thomas Übel reminds me that Kurt Gödel did not leave for the last time until 1940.

2. Herbert Feigl was both one of those listed as members of the Vienna Circle in the 1929 manifesto and one of the editors of the 1961 volume which represents a high water mark for Logical Empiricism in North America. Carnap and Reichenbach, followed by Hempel, were undoubtedly the intellectual leaders of Logical Empiricism in North America, but Feigl, more than anyone else, created the institutional basis for the movement.

3. The nature and circumstances of Reichenbach's appointment in Berlin were complicated. Apparently the original plan was to create for Reichenbach a position as "ausserordentlicher Professor für Naturphilosophie." This appointment was rejected by a faculty vote of 26 to 8, with 5 abstentions (Hecht and Hoffmann 1982, 659). This vote seems to have included members of the philosophy faculty. The reasons for rejecting Reichenbach included his liberal political views, as evidenced by his involvement with socialist student groups after World War I, but his outspoken anti-metaphysical philosophical positions, which privileged scientific theory over philosophical metaphysics, also played a role. In the end, Laue and other physicists secured for him an appointment in the physics faculty as "nichtbeamteter ausserordentlicher Professor" assigned to teach "die erkenntnistheoretischen Grundlagen der Physik." This appears to have been an untenured (nichtbeamtet) position. I thank Don Howard for helping me sort out this issue.

4. One possible but mistaken answer is that these works did not need to be translated because everyone who needed to read them could do so in the original German. World War II accelerated a disengagement with German culture that had begun with World War I. Before 1914, of course, the large numbers of German immigrants, particularly from around the years 1848 and 1871, insured that German language and culture were well represented in North America. By the end of World War II, the numbers of students in North America learning the German language had declined dramatically. Thus few students in the postwar period could read the founding documents of scientific philosophy in the original even if they had been so inclined, which most apparently were not.

5. In the case of Carnap, there is another possible explanation urged by Alan Richardson. It is simply that by 1936 Carnap regarded his earlier work as mistaken and thus not worth translating. Richardson cites an unpublished letter written by Carnap in 1938 expressing very negative views of the *Aufbau* to an inquiring Nelson Goodman. And it is, of course, very plausible that Feigl would have consulted with Carnap about what works should appear in his *Readings*. On the other hand, that a philosopher later regards an important earlier work as mistaken is not typically taken as grounds for later philosophers not to read it. The *Tractatus* again provides a clear contrast.

6. This applies doubly to the socialist rhetoric of Neurath's "Wissenschaftliche Weltauffassung" (1929), which would have been damaging to the movement in the America of the 1930s, 40s, and 50s.

7. This information comes from a conversation with Prof. Hempel in Princeton in the spring of 1992.

8. The truth of this claim depends, of course, on how one defines "most active areas of research." I am quite convinced that a count of articles or citations would support this claim, but do not have actual numbers to back up that conviction. For a relatively recent overview of the extensive literature on explanation, which demonstrates the enduring interest in the topic, see (Salmon 1989). No similar review exists for confirmation.

9. I owe the realization that the notion of explication appears in *Meaning and Necessity* to Thomas Ryckman. Michael Friedman reminded me that it appears also in the "Two Concepts" paper.

10. In his early expositions (1947, 1950), Carnap refers to C. H. Langford's discussion of G. E. Moore's notion of philosophical analysis.

11. In the Schilpp volume (1963), Carnap's attitude toward ordinary language analysis comes out clearly in his exchange with P. F. Strawson.

12. For an exemplar of this style of philosophy of science, see Israel Scheffler's *Anatomy of Inquiry* (1963).

13. See Friedman (1987, 1992) and Richardson (1990, 1992).

14. I learned this story from Andreas Kamlah, who knew neither of its origin nor of any evidence that it might be true.

15. This is the paper, by the way, which is reprinted in translation in Feigl and Brodbeck (1953).

16. This story too I owe to Andreas Kamlah, but this one seems to be well established.

17. Thomas Ryckman suggested to me the following elaboration of this point, based on a letter of Reichenbach's to Einstein in 1936 (see Ryckman 1996). Reichenbach had clashed with Hermann Weyl over relativity theory in the 1920s. By 1936, Weyl was a powerful figure in the American academic establishment, powerful enough to block any possibility of Reichenbach's obtaining a position at Princeton. Moreover, as became clear in Einstein's response to Reichenbach's contribution to his Schilpp volume (1949b), Einstein himself was critical of Reichenbach's understanding of the epistemological significance of relativity theory. It is a

reasonable inference that Reichenbach refused to write on space, time, and relativity for the *Encyclopedia* because he did not want these disagreements to become well known in America.

18. The foreword is dated Istanbul, August 1934.

19. Another version, with proper Kantian overtones, appears in Karl Popper's *Logik der Forschung* (1935). Popper's version, however, seems to have had little influence on philosophical thought in America before publication of the revised English edition of the *Logik* (Popper 1959).

20. One must here recall how vicious and personal the attacks on Einstein by Nazi sympathizers were in 1932–33 (Clark 1971). Einstein himself was in the United States when Hitler came to power and shortly thereafter resigned his posts from the safety of his temporary residence in Belgium. He did not return to Germany. Reichenbach, by contrast, remained in Berlin long enough to experience first hand the attacks on his patron.

21. For a good contemporary survey of the state of American philosophy in 1930, see Adams and Montague (1930). For a recent history of American philosophy up to 1930, see Wilson (1990).

22. This is not to deny that there continued to be individual scholars of considerable renown drawing inspiration from the pragmatists, particularly Peirce, whose collected works began appearing in 1931 (see Peirce 1931–58). These included Arthur Burks at Michigan, Ernest Nagel at Columbia, and both Nicholas Rescher and Wilfrid Sellars at Pittsburgh. Both Burks and Rescher published major works in the 1970s which exhibit the continuing influence of Peirce (Burks 1977; Rescher 1973a, 1973b, 1977).

23. The heart of Reichenbach's "pragmatic justification" of induction is the purely mathematical proof, utilizing the mathematical definition of the limit of an infinite sequence, that the "straight rule" must eventually lead to ever more accurate posits of the true limit—provided only that such a limit exists. But if no limit exists, no rule could lead us to it. So we have everything to gain and nothing to lose by following the rule. (Reichenbach 1949a, 469–82).

24. Reichenbach himself sounded this theme in the final paragraph of his contribution (1939, 192) to Dewey's Schilpp volume. There he wrote: "The early period of empiricism in which an all-round philosopher could dominate at the same time the fields of scientific method, of history of philosophy, of education and social philosophy, has passed. We enter into the second phase in which highly technical investigations form the indispensable instrument of research, splitting the philosophical campus into specialists of its various branches." In other words, Dewey's (and Pragmatism's) time has passed; we enter the time of Reichenbach (and Logical Empiricism).

25. How different this was from the situation in Germany following World War I, where the natural sciences were often identified with the militarism that had led to Germany's destruction. I wonder whether those who experienced both periods, including Carnap and Reichenbach, realized the irony of the difference.

26. Here I simply follow the biography in Schilpp and Hahn (1939).

27. Dewey's commitment to socialism, and his involvements in the social

movements of the 1930s and 1940s, are examined in the final chapters of Robert Westbrook's (1991) lucid biography.

28. John McCumber (1996) has explored connections between McCarthyism and the development of analytic philosophy following World War II. His analysis supports my speculations regarding the decline of Pragmatism.

29. McCumber (1996) reports that Carnap initially refused appointment at UCLA because of the loyalty oath in force in California in the early 1950s. Arthur Fine told me that letters in the Einstein papers show that Carnap similarly was reluctant to accept an appointment at Princeton because of discrimination there against Jews. (Reichenbach reportedly had no such compunctions.) These private acts of conscience, however laudable, do not alter the fact that the public face of Logical Empiricism exhibited mostly the logical analysis of scientific theories and methods.

30. Reichenbach's popular book *The Rise of Scientific Philosophy* (1951) may be an exception to the general lack of involvement by Logical Empiricists in social issues of the day. David Hollinger (1995) has recently placed this work within in a genre of works of the 1940s and 50s extolling the value of science for liberal democratic societies, and thus opposing the nationalistic anticommunism exemplified by McCarthyism. I recall Wesley Salmon once telling me that Reichenbach had written the book because he had promised his second wife, Maria, that he would write her a "best seller" after they got to America. This latter motivation is compatible with Hollinger's analysis of the role of the book in the culture of the time. It is also possible that Reichenbach himself had political motivations in writing the book but sought to conceal them from his American students.

31. The historian of science Mitchell Ash suggested to me a striking contrast between the scientific philosophers and another German immigrant group, the gestalt psychologists. The latter were generally unsuccessful in achieving academic positions that provided access to graduate students. See for example Ash (1995), Ash and Söllner (1996).

32. Nagel is especially worth studying in this regard since he was identified both with Pragmatism and later with Logical Empiricism. It is instructive to compare Cohen and Nagel (1934) with Nagel (1961). His best known students are identified more with Logical Empiricism than with Pragmatism, although one can find pragmatist sympathies in the work of both Isaac Levi and Patrick Suppes. Nagel himself recommended the study of Peirce's philosophy at the Fifth International Congress for the Unity of Science at Harvard in 1939 (Nagel 1940), and he defended naturalism as late as his 1954 presidential address for the Eastern Division of the American Philosophical Association (Nagel 1956).

33. Frank had been part of "The First Vienna Circle" in the years 1908–12. On Einstein's recommendation, he was appointed in 1913 to fill the position vacated by Einstein as Professor of Physics at the German University in Prague. He was a member of the later Vienna Circle and instrumental in bringing Carnap to Prague in 1931. He remained a part-time Professor of Physics and Philosophy at Harvard until retirement in the middle 1950s. He died in 1966.

34. In an unpublished talk for a plenary session at the 1991 meeting of the

Society for Social Studies of Science, Gerald Holton suggests that the founding of the program for Science, Technology, and Society at MIT in 1977 was also partly a result of Frank's endeavors. The then President of MIT, Jerome Wiesner, and his Provost, Walter Rosenblith, had both attended some meetings of the Institute for the Unity of Science, and shared some of Frank's visions for science. I thank Prof. Holton for providing me with a copy of this manuscript.

35. Consider arguably the most influential and prolific philosopher of science in the "third" generation, Bas van Fraassen. He was born during the war in the Netherlands and emigrated with his family to Canada while a teenager. He completed his Ph.D. with Grünbaum at Pittsburgh, thus establishing a lineage going directly back to Reichenbach by way of Hempel. Like Reichenbach, van Fraassen has combined general logical and methodological works (1980, 1989) with philosophical studies of both relativity theory (1970a) and quantum mechanics (1991).

# References

Adams, G. P., and W. P. Montague, eds. 1930. *Contemporary American Philosophy,* 2 vols. New York: Russell and Russell.
Armstrong, D. 1983. *What is a Law of Nature?* Cambridge: Cambridge University Press.
Aronson, J. L., R. Harré, and E. C. Way. 1995. *Realism Rescued: How Scientific Progress is Possible.* London: Duckworth.
Ash, M. G. 1995. *Gestalt Psychology in German Culture, 1890–1967: Holism and the Quest for Objectivity.* Cambridge: Cambridge University Press.
Ash, M. G., and A. Söllner, eds. 1996. *Forced Migration and Scientific Change: Emigré German-Speaking Scientists and Scholars after 1933.* Cambridge: Cambridge University Press.
Ayer, A. J. 1936. *Language, Truth, and Logic.* London: Gollancz.
Baltzer, W., and C. U. Moulines, eds. 1996. *Structuralist Theory of Science.* New York: Walter de Gruyter.
Barnes, B. 1974. *Scientific Knowledge and Sociological Theory.* London: Routledge & Kegan Paul.
———. 1982. *T. S. Kuhn and Social Science.* New York: Columbia University Press.
Barnes, B., and D. Edge, eds. 1982. *Science in Context.* Cambridge: MIT Press.
Barnes, B., and S. Shapin, eds. 1979. *Natural Order: Historical Studies of Scientific Culture.* Beverly Hills: Sage.
Berlin, B., and P. Kay. 1969. *Basic Color Terms: Their Universality and Evolution.* Berkeley: University of California Press.
Biagioli, M. 1993. *Galileo, Courtier: The Practice of Science in the Culture of Absolutism.* Chicago: University of Chicago Press.
Bijker, W., T. Pinch, and T. Hughes, eds. 1987. *The Social Construction of Techno-*

*logical Systems: New Directions in the Sociology and History of Technology.* Cambridge: MIT Press.
Bloor, D. 1976. *Knowledge and Social Imagery.* London: Routledge & Kegan Paul.
Boyd, R. 1973. Realism, Underdetermination, and a Causal Theory of Evidence. *Noûs* 7:1–12.
———. 1981. Scientific Realism and Naturalistic Epistemology. In *PSA 1980*, vol. 2, ed. P. D. Asquith and R. N. Giere, 613–62. East Lansing, MI: The Philosophy of Science Association.
Bradie, M. 1986. Assessing Evolutionary Epistemologies. *Biology and Philosophy* 1: 401–59.
Brannigan, A. 1981. *The Social Basis of Scientific Discoveries.* Cambridge: Cambridge University Press.
Brooke, J. H. 1991. *Science and Religion: Some Historical Perspectives.* New York: Cambridge University Press.
Bruner, J. S., J. J. Goodnow, and G. A. Austin. 1956. *A Study of Thinking.* New York: Wiley.
Burks, A. W. 1977. *Cause, Chance, and Reason.* Chicago: University of Chicago Press.
Callebaut, W., and R. Pinxten, eds. 1987. *Evolutionary Epistemology.* Dordrecht: Reidel.
Campbell, D. T. 1960. Blind Variation and Selective Retention in Creative Thought as in Other Knowledge Processes. *Psychological Review* 67:380–400.
———. 1974. Evolutionary Epistemology. In *The Philosophy of Karl Popper*, ed. P. A. Schilpp, 413–63. La Salle, IL: Open Court.
Carey, S. 1985. *Conceptual Change in Childhood.* Cambridge: MIT Press.
Carnap, R. 1928. *Der logische Aufbau der Welt.* Berlin: Weltkreis Verlag.
———. 1936–37. Testability and Meaning. *Philosophy of Science* 3:419–71, 4:1–40.
———. 1937. *The Logical Syntax of Language.* London: Routledge & Kegan Paul.
———. 1939. *Foundations of Logic and Mathematics.* Chicago: University of Chicago Press.
———. 1942. *Introduction to Semantics.* Cambridge: Harvard University Press.
———. 1945a. The Two Concepts of Probability. *Philosophy and Phenomenological Research* 5:513–32.
———. 1945b. On Inductive Logic. *Philosophy of Science* 12:72–97.
———. 1947. *Meaning and Necessity.* Chicago: University of Chicago Press.
———. 1949. Truth and Confirmation. In *Readings in Philosophical Analysis*, ed. H. Feigl and W. Sellars, 119–27. New York: Appleton-Century-Crofts.
———. 1950. *Logical Foundations of Probability.* Chicago: University of Chicago Press (2d ed. 1962).
———. 1963. Intellectual Autobiography. In *The Philosophy of Rudolf Carnap*, ed. P. A. Schilpp, 3–84. LaSalle, IL: Open Court.
———. 1967. *The Logical Construction of the World.* Tr. R. A. George. Berkeley: University of California Press.

Cartwright, N. D. 1983. *How the Laws of Physics Lie*. Oxford: Clarendon Press.
———. 1989. *Nature's Capacities and Their Measurement*. Oxford: Oxford University Press.
Chi, M. T. H., P. J. Feltovich, and R. Glaser. 1981. Categorization and Representation of Physics Problems by Experts and Novices. *Cognitive Science* 5: 121–52.
Churchland, P. M. 1979. *Scientific Realism and the Plasticity of Mind*. Cambridge: Cambridge University Press.
———. 1989. *A Neurocomputational Perspective: The Nature of Mind and the Structure of Science*. Cambridge: MIT Press.
Churchland, P. S. 1986. *Neurophilosophy: Toward a Unified Science of the Mind-Brain*. Cambridge: MIT Press.
Churchland, P. S., and Churchland, P. M. 1983. Stalking the Wild Epistemic Engine. *Noûs* 17: 5–18.
Clark, R. W. 1971. *Einstein: The Life and Times*. New York: World Publishing.
Coffa, J. A. 1973. *Foundations of Inductive Explanation*. Ann Arbor: University Microfilms.
Cohen, I. B. 1985. *Revolution in Science*. Cambridge: Harvard University Press.
Cohen, M. R., and E. Nagel. 1934. *An Introduction to Logic and Scientific Method*. New York: Harcourt.
Collins, H. M. 1985. *Changing Order: Replication and Induction in Scientific Practice*. London: Sage.
———, ed. 1982. *Sociology of Scientific Knowledge: A Source Book*. Bath: Bath University Press.
Collins, H. M., and T. J. Pinch. 1982. *Frames of Meaning*. London: Routledge & Kegan Paul.
Collins, H., and S. Yearley. 1992. Epistemological Chicken. In *Science as Practice and Culture*, ed. A. Pickering, 283–300. Chicago: University of Chicago Press.
Cox, A. 1969. Geomagnetic Reversals. *Science* 163: 237–45.
Cox, A., R. R. Doell, and G. B. Dalrymple. 1963. Geomagnetic Polarity Epochs and Pleistocene Geochronometry. *Nature* 198: 1049–51.
———. 1964. Reversals of the Earth's Magnetic Field. *Science* 144: 1537–43.
Desmond, A., and J. Moore. 1991. *Darwin*. London: Penguin.
Dewey, J. 1910. *The Influence of Darwin on Philosophy and Other Essays in Contemporary Thought*. New York: Henry Holt & Co.
———. 1930. *The Quest for Certainty*. London: Allen and Unwin.
———. 1938. *Logic: The Theory of Inquiry*. New York: Holt.
———. 1939. No Matter What Happens—Stay Out! *Common Sense* 8(3):11.
Doppelt, G. 1978. Kuhn's Epistemological Relativism: An Interpretation and Defense. *Inquiry* 21: 33–86.
Earman, J. 1986. *A Primer on Determinism*. Reidel: Dordrecht.
———. 1995. *Bangs, Crunches, Whimpers, and Shrieks: Singularities and Acausalities in Relativistic Spacetimes*. New York: Oxford University Press.
Einstein, A. 1934. What is the Theory of Relativity? In *Essays in Science*. Tr. E. Harris. New York: Philosophical Library.

Feigl, H. 1981. *Inquiries and Provocations: Selected Writings, 1929–1974.* Ed. R. S. Cohen. Boston: D. Reidel.

Feigl, H., and M. Brodbeck, eds. 1953. *Readings in the Philosophy of Science.* New York: Appleton-Century-Crofts.

Feigl, H., and G. Maxwell, eds. 1961. *Current Issues in the Philosophy of Science.* New York: Holt, Rinehart & Winston.

Feigl, H., and M. Scriven, eds. 1956. *The Foundations of Science and the Concepts of Psychology and Psychoanalysis,* Minnesota Studies in the Philosophy of Science, vol. I. Minneapolis: University of Minnesota Press.

Feigl, H., and W. Sellars, eds. 1949. *Readings in Philosophical Analysis.* New York: Appleton-Century-Crofts.

Feyerabend, P. K. 1962. Explanation, Reduction, and Empiricism. In *Scientific Explanation, Space, and Time,* Minnesota Studies in the Philosophy of Science, vol. III, ed. H. Feigl and G. Maxwell, 28–97. Minneapolis: University of Minnesota Press.

Fine, A. 1984. And Not Anti-Realism Either. *Noûs* 18:51–65.

———. 1986. *The Shaky Game: Einstein, Realism, and the Quantum Theory.* Chicago: University of Chicago Press (2d ed. 1996).

Frankel, H. 1982. The Development, Reception and Acceptance of the Vine-Matthews-Morley Hypothesis. *Historical Studies in the Physical Sciences* 13: 1–39.

———. 1987. The Continental Drift Debate. In *Scientific Controversies,* ed. H. T. Engelhardt Jr. and A. L. Caplan, 203–48. Cambridge: Cambridge University Press.

Franklin, A. 1986. *The Neglect of Experiment.* Cambridge: Cambridge University Press.

———. 1990. *Experiment, Right or Wrong.* Cambridge: Cambridge University Press.

Friedman, M. 1979. Truth and Confirmation. *The Journal of Philosophy* 76:361–82.

———. 1987. Carnap's *Aufbau* Reconsidered. *Noûs* 21:521–45.

———. 1991. The Re-evaluation of Logical Positivism. *Journal of Philosophy* 88: 505–19.

———. 1992. Epistemology in the *Aufbau. Synthese* 93:15–57.

Gähde, U. 1989. Bridge Structures and the Borderline Between the Internal and External History of Science. In *Imre Lakatos and Theories of Scientific Change,* ed. K. Gavroglu, Y. Goudaroulis, and P. Nicolacopoulos, 215–25. Dordrecht: Kluwer.

Galison, P. L. 1987. *How Experiments End.* Chicago: University of Chicago Press.

———. 1997. *Image and Logic: A Material Culture of Microphysics.* Chicago: University of Chicago Press.

Gay, P. 1969. *The Enlightenment: An Interpretation.* New York: Knopf.

Giere, R. N. 1973. History and Philosophy of Science: Intimate Relationship or Marriage of Convenience? *British Journal for the Philosophy of Science* 24: 282–97.

———. 1975. The Epistemological Roots of Scientific Knowledge. In *Induction,*

*Probability, and Confirmation,* Minnesota Studies in the Philosophy of Science, vol. VI, ed. G. Maxwell and R. M. Anderson, Jr., 212–61. Minneapolis: University of Minnesota Press.

———. 1979. Foundations of Probability and Statistical Inference. In *Current Research in Philosophy of Science,* ed. P.D. Asquith and H. E. Kyburg Jr., 503–33. East Lansing, MI: Philosophy of Science Association.

———. 1983. Testing Theoretical Hypotheses. In *Testing Scientific Theories,* Minnesota Studies in the Philosophy of Science, vol. X, ed. John Earman, 269–98. Minneapolis: University of Minnesota Press.

———. 1984. Toward a Unified Theory of Science. In *Science and Reality,* ed. J. T. Cushing, C. F. Delaney, and G. Gutting, 5–31. Notre Dame: University of Notre Dame Press.

———. 1985. Constructive Realism. In *Images of Science,* ed. P. M. Churchland and C. A. Hooker, 75–98. Chicago: University of Chicago Press.

———. 1988. *Explaining Science: A Cognitive Approach.* Chicago: University of Chicago Press.

———. 1989a. The Units of Analysis in Science Studies. In *The Cognitive Turn: Sociological and Psychological Perspectives on Science,* Sociology of the Sciences Yearbook, vol. XIII, ed. S. Fuller, M. DeMey, T. Shinn, and S. Woolgar, 3–11. Dordrecht: Reidel.

———. 1989b. Scientific Rationality as Instrumental Rationality. *Studies in History and Philosophy of Science* 20:377–84.

———. 1989c. What Does Explanatory Coherence Explain? *Behavioral and Brain Sciences* 12:475–76.

———. 1991. Implications of the Cognitive Sciences for the Philosophy of Science. In *PSA 90,* vol. 1, ed. A. Fine, M. Forbes, and L. Wessels, 419–30. East Lansing, MI: The Philosophy of Science Association.

———. 1994a. The Cognitive Structure of Scientific Theories. *Philosophy of Science* 61:276–96.

———. 1994b. No Representation without Representation. *Biology and Philosophy* 9:113–20.

———. 1995. Science Without Laws of Nature. In *Laws of Nature,* ed. F. Weinert, 120–38. Hawthorne, NY: Walter de Gruyter.

———. 1996a. Visual Models and Scientific Judgment. In *Picturing Knowledge: Historical and Philosophical Problems Concerning the Use of Art in Science,* ed. B. S. Baigrie, 269–302. Toronto: University of Toronto Press.

———. 1996b. The Feminism Question in the Philosophy of Science. In *Feminism, Science, and the Philosophy of Science,* ed. L. H. Nelson and J. Nelson, 3–15. Boston: Kluwer.

———. 1996c. From *Wissenschaftliche Philosophie* to Philosophy of Science. In *Origins of Logical Empiricism,* Minnesota Studies in the Philosophy of Science, vol. XVI, ed. R. N. Giere and A. Richardson, 335–54. Minneapolis: University of Minnesota Press.

———, ed. 1992. *Cognitive Models of Science.* Minnesota Studies in the Philosophy of Science, vol. XV. Minneapolis: University of Minnesota Press.

Giere, R. N., and A. Richardson, eds. 1996. *Origins of Logical Empiricism,* Minnesota Studies in the Philosophy of Science, vol. XVI. Minneapolis: University of Minnesota Press.

Gilligan, C. 1982. *In a Different Voice.* Cambridge: Harvard University Press.

Glen, W. 1982. *The Road to Jaramillo.* Stanford, CA: Stanford University Press.

Gooding, D. 1990. *Experiment and the Making of Meaning.* Dordrecht: Kluwer.

Gooding, D., T. Pinch, and S. Schaffer. 1989. *The Uses of Experiment: Studies in the Natural Sciences.* Cambridge: Cambridge University Press.

Gould, S. J., and N. Eldredge. 1977. Punctuated Equilibria: The Tempo and Mode of Evolution Reconsidered. *Paleobiology* 3:115–51.

Gross, P. R., and N. Levitt. 1994. *Higher Superstition: The Academic Left and Its Quarrels with Science.* Baltimore: The Johns Hopkins University Press.

Gross, P. R., N. Levitt, and M. W. Lewis, eds. 1996. *The Flight From Science and Reason.* New York: The New York Academy of Sciences.

Grünbaum, A., and W. Salmon, eds. 1988. *The Limitations of Deductivism.* Berkeley: University of California Press.

Hacking, I. 1968. One Problem about Induction. In *The Problem of Inductive Logic,* ed. I. Lakatos, 44–58. Amsterdam: North-Holland.

———. 1983. *Representing and Intervening.* Cambridge: Cambridge University Press.

Hahlweg, K., and C. A. Hooker. 1989. *Issues in Evolutionary Epistemology.* Albany, NY: SUNY Press.

Hale, B., and C. Wright. 1997. Putnam's Model-Theoretic Argument Against Metaphysical Realism. In *A Companion to the Philosophy of Language,* ed. B. Hale and C. Wright, 427–57. Oxford: Blackwell.

Hanson, N. R. 1958. The Logic of Discovery. *Journal of Philosophy* 55:1073–89.

———. 1961. Is There a Logic of Discovery? In *Current Issues in the Philosophy of Science,* ed. H. Feigl and G. Maxwell, 20–35. New York: Holt, Rinehart & Winston.

Haraway, D. J. 1991. *Simians, Cyborgs, and Women.* New York: Routledge.

Harding, S. 1986. *The Science Question in Feminism.* Ithaca, NY: Cornell University Press.

Harré, R. 1987. *Varieties of Realism.* Oxford: Blackwell.

———. 1993. *Laws of Nature.* London: Duckworth.

Hatfield, G. 1990. *The Natural and the Normative: Theories of Perception from Kant to Helmholtz.* Cambridge: MIT Press.

Healey, R. 1989. *The Philosophy of Quantum Mechanics: An Interactive Interpretation.* Cambridge: Cambridge University Press.

Hecht, H., and D. Hoffmann. 1982. Die Berufung Hans Reichenbachs an die Berliner Universitaet. *Deutsche Zeitschrift für Philosophie* 30:651–62.

Hempel, C. G. 1945. Studies in the Logic of Confirmation. *Mind* 54:1–26, 97–121.

———. 1965. *Aspects of Scientific Explanation.* New York: Free Press.

———. 1983. Valuation and Objectivity in Science. In *Physics, Philosophy and Psychoanalysis,* ed. R. S. Cohen and L. Laudan, 73–100. Dordrecht: D. Reidel.

———. 1988. Provisos: A Problem Concerning the Inferential Function of Scientific Theories. In *The Limitations of Deductivism,* ed. A. Grünbaum and W. Salmon. Berkeley: University of California Press.

Hempel, C. G., and P. Oppenheim. 1948. Studies in the Logic of Explanation. *Philosophy of Science* 15:135–75.

Hess, H. H. 1962. History of Ocean Basins. In *Petrologic Studies,* ed. A. E. J. Engel, H. L. James, and B. F. Leonard, 599–620. Boulder, CO: The Geological Society of America.

Hollinger, D. A. 1983. The Defense of Democracy and Robert K. Merton's Formulation of the Scientific Ethos. *Knowledge and Society* 4:1–15.

———. 1995. Science as a Weapon in *Kulturkämpfe* in the United States during and after World War II. *Isis* 86:440–54.

Holmes, A. 1930. Radioactivity and Earth Movements. *Transactions of the Geological Society of Glasgow* 18:559–606.

———. 1944. *Principles of Physical Geology.* London: Nelson.

Holton, G. 1992. Ernst Mach and the Fortunes of Positivism in America. *Isis* 83:27–60.

———. 1993. From the Vienna Circle to Harvard Square: The Americanization of a European World Conception. In *Scientific Philosophy: Origins and Developments,* ed. F. Stadler, 47–73. Boston: Kluwer.

———. 1995. On the Vienna Circle in Exile: An Eyewitness Report. In *The Foundational Debate,* ed. E. Kohler, W. Schimanovich, and F. Stadler, 269–92. Boston: Kluwer.

Hooker, C. A. 1987. *A Realistic Theory of Science.* Albany: SUNY Press.

———. 1995. *Reason, Regulation, and Realism: Toward a Regulatory Systems Theory of Reason and Evolutionary Epistemology.* Albany: SUNY Press.

Howson, C., and P. Urbach. 1989. *Scientific Reasoning: The Bayesian Approach.* La Salle, IL: Open Court.

Hoyningen-Huene. P. 1987. Context of Discovery and Context of Justification. *Studies in the History and Philosophy of Science* 18:501–15.

———. 1993. *Reconstructing Scientific Revolutions: Thomas S. Kuhn's Philosophy of Science.* Chicago: University of Chicago Press.

Hughes, T. P. 1983. *Networks of Power: Electrification in Western Society, 1880–1930.* Baltimore: Johns Hopkins University Press.

———. 1989. *American Genesis: A Century of Innovation and Technological Enthusiasm, 1870–1970.* New York: Penguin Books.

Hull, D. 1988. *Science as a Process: An Evolutionary Account of the Social and Conceptual Development of Science.* Chicago: University of Chicago Press.

Jacob, M. C. 1976. *The Newtonians and the English Revolution: 1689–1720.* Ithaca, NY: Cornell University Press.

James, W. 1907. *Pragmatism.* New York: Longmans, Green, and Co.

Janik, A., and S. Toulmin. 1973. *Wittgenstein's Vienna.* New York: Simon & Schuster.

Jeffrey, R. C. 1973. Carnap's Inductive Logic. *Synthese* 25:299–306.

———. 1965. *The Logic of Decision.* New York: McGraw-Hill (2d ed. Chicago: University of Chicago Press, 1983).
———. 1985. Probability and the Art of Judgment. In *Observation, Experiment, and Hypothesis in Modern Physical Science,* ed. P. Achinstein and O. Hannaway, 95–126. Cambridge: MIT Press.
Kahneman, D., P. Slovic, and A. Tversky, eds. 1982. *Judgment Under Uncertainty: Heuristics and Biases.* Cambridge: Cambridge University Press.
Keil, F. C. 1989. *Concepts, Kinds, and Cognitive Development.* Cambridge, MA: MIT Press.
Keller, E. F. 1983. *A Feeling for the Organism.* New York: W. H. Freeman.
———. 1985. *Reflections on Gender and Science.* New Haven: Yale University Press.
———. 1992. *Secrets of Life, Secrets of Death: Essays on Language, Gender, and Science.* New York: Routledge.
———. 1995. *Refiguring Life: Metaphors of Twentieth-Century Biology.* New York: Columbia University Press.
Keynes, J. M. 1921. *A Treatise on Probability.* London: Macmillan.
Kitcher, P. 1983. *Abusing Science.* Cambridge: MIT Press.
———. 1993. *The Advancement of Science.* Oxford: Oxford University Press.
Knorr-Cetina, K. D. 1981. *The Manufacture of Knowledge.* Oxford: Pergamon Press.
Knorr-Cetina, K. D., and M. Mulkay, eds. 1983. *Science Observed.* Hollywood, CA: Sage.
Kohlberg, L. 1973. *Collected Papers on Moral Development and Moral Education.* Cambridge: Harvard University Press.
Kripke, S. 1972. Naming and Necessity. In *The Semantics of Natural Language,* ed. G. Harman and D. Davidson. Dordrecht: Reidel.
Kuhn, T. S. 1962. *The Structure of Scientific Revolutions.* Chicago: University of Chicago Press (2d ed. 1970).
———. 1974. Second Thoughts on Paradigms. In *The Structure of Scientific Theories,* ed. F. Suppe, 459–82. Urbana: University of Illinois Press.
———. 1977. *The Essential Tension.* Chicago: University of Chicago Press.
Kulkarni, D., and H. Simon. 1988. The Processes of Scientific Discovery: The Strategy of Experimentation. *Cognitive Science* 12:139–75.
Kusch, M. 1995. *Psychologism: A Case Study in the Sociology of Philosophical Knowledge.* London: Routledge.
Lakatos, I. 1970. Falsification and the Methodology of Scientific Research Programmes. In *Criticism and the Growth of Knowledge,* ed. I. Lakatos and A. Musgrave, 91–195. Cambridge: Cambridge University Press.
———. 1971. History of Science and Its Rational Reconstructions. In *PSA 1970,* Boston Studies in the Philosophy of Science, vol. 8, ed. R. S. Cohen and R. C. Buck, 91–135. Dordrecht: D. Reidel.
———. 1978. *Philosophical Papers.* Ed. J. Worrall and G. Currie. 2 vols. Cambridge: Cambridge University Press.
Lakoff, G. 1987. *Women, Fire, and Dangerous Things: What Categories Reveal about the Mind.* Chicago: University of Chicago Press.

Langley, P., H. A. Simon, G. L. Bradshaw, and J. M. Zytkow. 1987. *Scientific Discovery.* Cambridge: MIT Press.

Larkin, J. H. 1985. Understanding, Problem Representation, and Skill in Physics. In *Thinking and Learning Skills,* ed. S. F. Chipman, J. W. Segal, and R. Glaser, 141–59. Hillsdale, NJ: Erlbaum.

Larkin, J. H., J. McDermott, D. P. Simon, and H. A. Simon. 1980. Expert and Novice Performance in Solving Physics Problems. *Science* 208:1335–42.

Latour, B. 1986. Visualization and Cognition: Thinking with Eyes and Hands. *Knowledge and Society: Studies in the Sociology of Culture, Past and Present* 6: 1–40.

———. 1987. *Science in Action.* Cambridge: Harvard University Press.

———. 1989. *The Pasteurization of France.* Cambridge: Harvard University Press.

———. 1993. *We Have Never Been Modern.* Cambridge: Harvard University Press.

Latour, B., and S. Woolgar. 1979. *Laboratory Life.* Beverly Hills: Sage. (2d ed. Princeton: Princeton University Press, 1986.)

Laudan, L. 1977. *Progress and Its Problems.* Berkeley: University of California Press.

———. 1981. A Confutation of Convergent Realism. *Philosophy of Science* 48: 19–48.

———. 1984. *Science and Values: The Aims of Science and Their Role in Scientific Debate.* Berkeley: University of California Press.

———. 1987. Progress or Rationality? The Prospects for Normative Naturalism. *American Philosophical Quarterly* 24:19–31.

Laudan, R. 1981. The Recent Revolution in Geology and Kuhn's Theory of Scientific Change. In *PSA 1978,* vol. 2, ed. P. D. Asquith and I. Hacking, 227–39. East Lansing, MI: Philosophy of Science Association.

———, ed. 1984. *The Nature of Technological Knowledge: Are Models of Scientific Change Relevant?* Dordrecht: Reidel.

Le Grand, H. E. 1988. *Drifting Continents and Shifting Theories.* Cambridge: Cambridge University Press.

———. 1990. Is a Picture Worth a Thousand Experiments? In *Experimental Inquiries: Historical, Philosophical and Social Studies of Experimentation in Science,* ed. H. E. Le Grand, 241–70. Boston: Kluwer.

Longino, H. E. 1990. *Science as Social Knowledge.* Princeton: Princeton University Press.

Lorentz, H. A., A. Einstein, H. Minkowski, and H. Weyl. 1923. *The Principle of Relativity: A Collection of Original Memoirs on the Special and General Theories of Relativity.* London: Methuen.

Lynch, M., and S. Woolgar, eds. 1990. *Representation in Scientific Practice.* Cambridge: MIT Press.

MacKenzie, D. A. 1981. *Statistics in Britain: 1865–1930.* Edinburgh: Edinburgh University Press.

———. 1990. *Inventing Accuracy: A Historical Sociology of Nuclear Missile Guidance.* Cambridge: MIT Press.

Mackenzie, D. A., and J. Wajcman, eds. 1985. *The Social Shaping of Technology.* Milton Keynes: Open University Press.

Marion, J. B. 1970. *Classical Dynamics*. 2d ed. New York: Academic Press.
Marvin, U. B. 1973. *Continental Drift: The Evolution of a Concept*. Washington, D. C.: Smithsonian Institution Press.
Maxwell, G. 1962. The Ontological Status of Theoretical Entities. In *Scientific Explanation, Space, and Time*, Minnesota Studies in the Philosophy of Science, vol. III, ed. H. Feigl and G. Maxwell, 3–27. Minneapolis: University of Minnesota Press.
Mayo, D. 1996. *Error and the Growth of Experimental Knowledge*. Chicago: University of Chicago Press.
McCumber, John. 1996. Time in the Ditch: American Philosophy and the McCarthy Era. *Diacritics* 26:33–49.
McGrayne, S. B. 1993. *Nobel Prize Women in Science*. Secaucus, N. J.: Carol Publishing Group.
McKinsey, J. C. C., A. C. Sugar, and P. Suppes. 1953. Axiomatic Foundations of Classical Particle Mechanics. *Journal of Rational Mechanics and Analysis* 2: 253–72.
McMullin, E. 1970. The History and Philosophy of Science: A Taxonomy. In *Historical and Philosophical Perspectives of Science*, Minnesota Studies in the Philosophy of Science, vol. V, ed. R. Stuewer, 12–67. Minneapolis: University of Minnesota Press.
Medin, D. 1989. Concepts and Conceptual Structure. *American Psychologist* 44: 1469–81.
Megill, A., ed. 1994. *Rethinking Objectivity*. Durham: Duke University Press.
Merton, R. K. 1973. *The Sociology of Science*, ed. N. Storer. New York: Free Press.
Milton, J. R. 1981. The Origin and Development of the Concept of the "Laws of Nature." *Archives Européennes de Sociologie* 22:173–95.
———. 1998. Laws of Nature. In *The Cambridge History of Seventeenth-Century Philosophy*, ed. D. Garber and M. Ayers, 680–701. Cambridge: Cambridge University Press.
Morris, C. 1937. *Logical Positivism, Pragmatism, and Scientific Empiricism*. Paris: Hermann et Cie.
Murphy, G. L., and D. L. Medin. 1985. The Role of Theories in Conceptual Coherence. *Psychological Review* 92:289–316.
Nagel, E. 1939. *Principles of the Theory of Probability*. Chicago: University of Chicago Press.
———. 1940. Charles S. Peirce, Pioneer of Modern Empiricism. *Philosophy of Science* 7:69–80.
———. 1956. Naturalism Reconsidered. In *Logic without Metaphysics*, 3–18. Glencoe, IL: Free Press.
———. 1961. *The Structure of Science*. New York: Harcourt, Brace, and World.
Needham, J. 1951. *The Grand Titration: Science and Society in East and West*. London: George Allen and Unwin.
Neurath, O. 1973 [1929]. The Scientific Conception of the World: The Vienna Circle [Wissenschaftliche Weltauffassung: Der Wiener Kreis]. In *Empiricism*

*and Sociology,* ed. M. Neurath and R. S. Cohen, 299–319. Dordrecht: D. Reidel.

Neurath, O., R. Carnap, and C. Morris, eds. 1955. *Foundations of the Unity of Science,* vol. 1. Chicago: University of Chicago Press.

Nickles, T., ed. 1980. *Scientific Discovery, Logic, and Rationality.* Dordrecht: Reidel.

Nisbett, R., and L. Ross. 1980. *Human Inference: Strategies and Shortcomings of Social Judgment.* Englewood Cliffs: Prentice Hall.

O'Keefe, J., and Nadel, L. 1978. *The Hippocampus as a Cognitive Map.* Oxford: Clarendon Press.

Oakley, F. 1961. Christian Theology and the Newtonian Science: The Rise of the Concept of the Laws of Nature. *Church History* 30:433–57. [Reprinted in (O'Connor and Oakley 1969).]

Oakley, R. 1984. *Omnipotence, Covenant, and Order.* Ithaca, New York: Cornell University Press.

O'Connor, D., and F. Oakley, eds.1969. *Creation: The Impact of an Idea.* New York: Charles Scribner's Sons.

Olby, R. 1974. *The Path to the Double Helix.* Seattle: University of Washington Press.

Opdyke, N. D., B. P. Glass, J. D. Hays, and J. H. Foster. 1966. Paleomagnetic Study of Antarctic Deep-Sea Cores. *Science* 154:349–57.

Peirce, C. S. 1931–58. *Collected Works.* Ed. C. Hartshorne, P. Weiss, and A. Burks. 8 vols. Cambridge: Harvard University Press.

Pellionisz, A., and Llinas, R. 1982. Space-Time Representation in the Brain: The Cerebellum as a Predictive Space-Time Metric Tensor. *Neuroscience* 7: 2949–70.

Piaget, J. 1954. *The Construction of Reality in the Child.* New York: Basic Books.

Pickering, A. 1984. *Constructing Quarks: A Sociological History of Particle Physics.* Chicago: University of Chicago Press.

———. 1995. *The Mangle of Practice: Time, Agency, and Science.* Chicago: University of Chicago Press.

———, ed. 1992. *Science as Practice and Culture.* Chicago: University of Chicago Press.

Pinch. T. 1986. *Confronting Nature: The Sociology of Solar Neutrino Detection.* Dordrecht: D. Reidel.

Pitman, W. C. III, and J. P. Heirtzler. 1966. Magnetic Anomalies over the Pacific-Antarctic Ridge. *Science* 154:1164–71.

Plantinga, A. 1991. When Faith and Reason Clash: Evolution and the Bible. *Christian Scholar's Review* 21:8–32.

———. 1993. *Warrant and Proper Function.* Oxford: Oxford University Press.

———. 1995. Methodological Naturalism. In *Facets of Faith and Science,* ed. J. van der Meer, 177–221. Lanham, MD: University Press of America.

———. 1996. Science: Augustinian or Duhemian. *Faith and Philosophy* 13: 368–94.

Polanyi, M. 1958. *Personal Knowledge.* London: Routledge & Kegan Paul.

Popper, K. R. 1935. *Logik der Forschung: Zur Erkenntnistheorie der Modernen Naturwissenschaft.* Vienna: Julius Springer.

———. 1959. *The Logic of Scientific Discovery.* London: Hutchinson.

———. 1972. *Objective Knowledge.* Oxford: Clarendon Press.

———. 1974. Intellectual Autobiography. In *The Philosophy of Karl Popper,* 2 vols., ed. P. A. Schilpp. La Salle: Open Court.

Putnam, H. 1975a. *Mathematics, Matter and Method.* Cambridge: Cambridge University Press.

———. 1975b. The Meaning of Meaning. In *Language, Mind, and Knowledge,* Minnesota Studies in the Philosophy of Science, vol. VII, ed. K. Gunderson. Minneapolis: University of Minnesota Press.

———. 1978. *Meaning and the Moral Sciences.* London: Routledge and Kegan Paul.

———. 1981. *Reason, Truth, and History.* Cambridge: Cambridge University Press.

———. 1982. Why Reason Can't Be Naturalized. *Synthese* 52:3–23.

Putnam, H., and Oppenheim, P. 1958. Unity of Science as a Working Hypothesis. In *Concepts, Theories, and the Mind-Body Problem,* Minnesota Studies in the Philosophy of Science, vol. II, ed. H. Feigl, M. Scriven, and G. Maxwell, 3–36. Minneapolis: University of Minnesota Press.

Quine. W. V. O. 1969. Epistemology Naturalized. In *Ontological Relativity and Other Essays,* 69–90. New York: Columbia University Press.

Reichenbach, H. 1920. *Relativitätstheorie und Erkenntnis Apriori.* Berlin: Springer.

———. 1924. *Axiomatik der Relativistischen Raum-Zeit-Lehre.* Braunschweig: Vieweg.

———. 1928. *Philosophie der Raum-Zeit-Lehre.* Berlin: Walter de Gruyter.

———. 1933. Die logischen Grundlagen des Wahrscheinlichkeits-begriffs. *Erkenntnis* 3:401–25.

———. 1935. *Wahrscheinlichkeitslehre.* Leiden: A. W. Sijthoff.

———. 1936. Logistic Empiricism in Germany and the Present State of Its Problems. *The Journal of Philosophy* 33:141–60.

———. 1938. *Experience and Prediction.* Chicago: University of Chicago Press.

———. 1939. Dewey's Theory of Science. In *The Philosophy of John Dewey,* ed. P. A. Schilpp and L. E. Hahn, 157–92. La Salle, IL: Open Court.

———. 1944. *Philosophical Foundations of Quantum Mechanics.* Berkeley: University of California Press.

———. 1947. *Elements of Symbolic Logic.* New York: Macmillan.

———. 1949a. *The Theory of Probability.* Berkeley: University of California Press.

———. 1949b. The Philosophical Significance of the Theory of Relativity. In *Albert Einstein: Philosopher-Scientist,* ed. P. A. Schilpp, 287–311. Evanston, IL: Open Court.

———. 1951. *The Rise of Scientific Philosophy.* Berkeley: University of California Press.

———. 1954. *Nomological Statements and Admissible Operations.* Amsterdam: North-Holland.

———. 1956. *The Direction of Time*. Ed. M. Reichenbach. Berkeley: University of California Press.

———. 1958. *The Philosophy of Space and Time*. Tr. M. Reichenbach and J. Freund. New York: Dover.

Rescher, N. 1973a. *Conceptual Idealism*. Oxford: Basil Blackwell.

———. 1973b. *The Primacy of Practice*. Oxford: Basil Blackwell.

———. 1977. *Methodological Pragmatism*. New York: New York University Press.

Richardson, A. 1990. How Not to Russell Carnap's *Aufbau*. In *PSA 1990*, vol. 1, ed. A. Fine, M. Forbes, and L. Wessels, 3–14. East Lansing, MI: Philosophy of Science Association.

———. 1992. Logical Idealism and Carnap's Construction of the World. *Synthese* 93:59–92.

Rorty, R. 1979. *Philosophy and the Mirror of Nature*. Princeton: Princeton University Press.

Rosch, E. 1973a. Natural Categories. *Cognitive Psychology* 4:328–50.

———. 1973b. On the Internal Structure of Perceptual and Semantic Categories. In *Cognitive Development and the Acquisition of Language*, ed. T. E. Moore. New York: Academic Press.

———. 1978. Principles of Categorization. In *Cognition and Categorization*, ed. E. Rosch and B. B. Lloyd. Hillsdale, NJ: Erlbaum.

Rosch, E., and C. B. Mervis. 1975. Family Resemblance Studies in the Internal Structure of Categories. *Cognitive Psychology* 7:573–605.

Rosch, E., et al. 1976. Basic Objects in Natural Categories. *Cognitive Psychology* 8: 382–439.

Ross, Andrew, ed. 1996. *Science Wars*. Durham: Duke University Press,.

Roth, P., and R. Barrett. 1990. Deconstructing Quarks. *Social Studies of Science* 20: 579–632.

Ruby, J. E. 1986. The Origins of Scientific "Law." *Journal of the History of Ideas* 46: 341–59. [Reprinted in (Weinert 1995).]

Ruse, M. 1981. What Kind of a Revolution Occurred in Geology? In *PSA 1978*, vol. 2, ed. P. D. Asquith and I. Hacking, 240–73. East Lansing, MI: The Philosophy of Science Association.

———. 1982. *Darwinism Defended*. Reading Mass.: Addison-Wesley.

———. 1986. *Taking Darwin Seriously*. Dordrecht: Reidel.

Russell, B. 1912. *The Problems of Philosophy*. Oxford: Oxford University Press.

———. 1914. *Our Knowledge of the External World*. London: Allen and Unwin.

———. 1939. Dewey's New *Logic*. In *The Philosophy of John Dewey*, ed. P. A. Schilpp, 135–56. La Salle, IL: Open Court.

Ryckman, T. 1996. Einstein *Agonistes:* Weyl and Reichenbach on Geometry and GTR. In *Origins of Logical Empiricism*, ed. R. N. Giere and A. Richardson, Minnesota Studies in the Philosophy of Science, vol. XVI, 165–209. Minneapolis: University of Minnesota Press.

Sacks, O. W. 1996. *The Island of the Colorblind and Cycad Island*. New York: Alfred A. Knopf.

Salmon, W. C. 1984. *Scientific Explanation and the Causal Structure of the World.* Princeton: Princeton University Press.

———. 1989. *Four Decades of Scientific Explanation.* Minneapolis: University of Minnesota Press.

Sargent, Rose-Mary. 1995. *The Diffident Naturalist: Robert Boyle and the Philosophy of Experiment.* Chicago: University of Chicago Press.

Scheffler, I. 1963. *The Anatomy of Inquiry: Philosophical Studies in the Theory of Science.* New York: Knopf.

———. 1967. *Science and Subjectivity.* New York: Bobbs-Merrill.

Scheman, Naomi. 1993. *Engenderings: Constructions of Knowledge, Authority, and Privilege.* New York: Routledge.

Schilpp, P. A., ed. 1949. *Albert Einstein: Philosopher-Scientist.* New York: Tudor.

———, ed. 1963. *The Philosophy of Rudolf Carnap.* LaSalle, IL: Open Court.

Schilpp, P. A., and L. E. Hahn, eds. 1939. *The Philosophy of John Dewey.* La Salle, IL: Open Court.

Schlick, M. 1925. *Allgemeine Erkenntnislehre.* 2d ed. Berlin: Springer.

———. 1936. Meaning and Verification, *Philosophical Review* 45:339–69.

———. 1975. *General Theory of Knowledge.* Tr. A. E. Blumberg. New York: Springer.

Scriven, M. 1961. The Key Property of Physical Laws—Inaccuracy. In *Current Issues in the Philosophy of Science,* ed. H. Feigl and G. Maxwell, 91–101. New York: Holt, Rinehart & Winston.

Shapere, D. 1964. The Structure of Scientific Revolutions. *Philosophical Review* 73: 383–94.

———. 1982. The Concept of Observation in Science and Philosophy. *Philosophy of Science* 49:485–525.

———. 1984. *Reason and the Search for Knowledge.* Dordrecht: Reidel.

Shapin, S. 1975. Phrenological Knowledge and the Social Structure of Early Nineteenth-Century Edinburgh. *Annals of Science* 32:219–43.

———. 1979. The Politics of Observation: Cerebral Anatomy and Social Interests in the Edinburgh Phrenology Disputes. In *On the Margins of Science: The Social Construction of Rejected Knowledge,* Sociological Review Monograph No. 27, ed. R. Wallis, 139–78. Keele: University of Keele Press.

———. 1982. History of Science and its Sociological Reconstructions. *History of Science* 20:157–211.

Shimony, A. 1971. Perception from an Evolutionary Point of View. *The Journal of Philosophy* 67:571–83.

———. 1981. Integral Epistemology. In *Scientific Inquiry and the Social Sciences,* ed. M. B. Brewer and B. E. Collins, 98–123. San Francisco: Jossey-Bass.

———. 1993. *Search for a Naturalistic World View.* 2 vols. Cambridge: Cambridge University Press.

Simon, H. A. 1954. The Axiomatization of Classical Mechanics. *Philosophy of Science* 21:340–43.

———. 1957. *Models of Man.* New York: Wiley.

———. 1972. Theories of Bounded Rationality. In *Decision and Organization,* ed. R. Radner and C. B. McGuire, 161–76. Amsterdam: North-Holland.

———. 1978. On the Forms of Mental Representation. In *Perception and Cognition: Issues in the Foundations of Psychology,* Minnesota Studies in the Philosophy of Science, vol. IX, ed. C. W. Savage, 3–18. Minneapolis: University of Minnesota Press.

Smith, E. E. 1990. Categorization. In *Thinking: An Invitation to Cognitive Science,* vol. 3. ed. D. N. Osherson and E. E. Smith, 33–53. Cambridge: MIT Press.

Smith, E. E., and D. L. Medin. 1981. *Categories and Concepts.* Cambridge, MA: Harvard University Press.

Sneed, J. D. 1971. *The Logical Structure of Mathematical Physics.* Dordrecht: D. Reidel.

Sober, E. 1984. *The Nature of Selection.* Cambridge: MIT Press.

Sokal, A. D. 1996a. Transgressing the Boundaries: Toward a Transformative Hermeneutics of Quantum Gravity. *Social Text* 46/47:217–52.

———. 1996b. A Physicist Experiments with Cultural Studies. *Lingua Franca* May/June: 62–64.

Stegmüller, W. 1976. *The Structure and Dynamics of Theories.* New York: Springer.

———. 1979. *The Structuralist View of Theories.* New York: Springer.

Steinle, F. 1995. The Amalgamation of a Concept—Laws of Nature in the New Sciences. In *Laws of Nature,* ed. F. Weinert, 316–68. New York: Walter de Gruyter.

Stewart, J. A. 1990. *Drifting Continents and Colliding Paradigms: Perspectives on the Geoscience Revolution.* Bloomington, IN: Indiana University Press.

Stroud, B. 1996. The Charm of Naturalism. *Proceedings and Addresses of the American Philosophical Association* 70:43–54.

Suppe, F. 1972. What's Wrong with the Received View on the Structure of Scientific Theories? *Philosophy of Science* 39:1–19.

———. 1973. Theories, Their Formulations, and the Operational Imperative. *Synthese* 25:129–64.

———. 1989. *The Semantic Conception of Theories and Scientific Realism.* Urbana, IL: University of Illinois Press.

Suppes, P. 1979. The Role of Formal Methods in the Philosophy of Science. In *Current Research in Philosophy in Science,* ed. P. D. Asquith and H. E. Kyburg, Jr., 16–27. East Lansing, MI: The Philosophy of Science Association.

Swartz, N. 1985. *The Concept of Physical Law.* Cambridge: Cambridge University Press.

Thagard, P. 1991. *Conceptual Revolutions.* Princeton: Princeton University Press.

Toulmin, S. 1953. *The Philosophy of Science.* London: Hutchinson.

———. 1961. *Foresight and Understanding.* New York: Harper & Row.

———. 1972. *Human Knowledge.* Princeton: Princeton University Press.

———. 1990. *Cosmopolis: The Hidden Agenda of Modernity.* New York: Free Press.

van der Gracht, W. A. J. M. van Waterschoot, et al. 1928. *Theory of Continental Drift.* Tulsa, OK: American Association of Petroleum Geologists.

van Fraassen, B. C. 1970a. *An Introduction to the Philosophy of Time and Space*. New York: Random House.

———. 1970b. On the Extension of Beth's Semantics of Physical Theories. *Philosophy of Science* 37: 325–39.

———. 1980. *The Scientific Image*. Oxford: Oxford University Press.

———. 1981. Theory Construction and Experiment: An Empiricist View. In *PSA 1980*, vol. 2, ed. P. D. Asquith and R. N. Giere, 663–78. East Lansing, MI: The Philosophy of Science Association.

———. 1985. Empiricism in the Philosophy of Science. In *Images of Science*, ed. P. M. Churchland and C. A. Hooker, 245–308. Chicago: University of Chicago Press.

———. 1989. *Laws and Symmetry*. Oxford: Oxford University Press.

———. 1991. *Quantum Mechanics: An Empiricist View*. Oxford: Oxford University Press.

———. 1993. Against Naturalized Epistemology. In *On Quine*, ed. P. Leonardi and M. Santambrogio, 68–88. Cambridge: Cambridge University Press.

Vine, F. J., and D. H. Matthews. 1963. Magnetic Anomalies over Oceanic Ridges. *Nature* 199:947–49.

Vine, F. J., and J. T. Wilson. 1965. Magnetic Anomalies over a Young Oceanic Ridge off Vancouver Island. *Science* 150:485–89.

Walker, L. J. 1984. Sex Differences in the Development of Moral Reasoning: A Critical Review. *Child Development* 55:677–91.

Wallner, F. 1990. *Acht Vorlesungen über den Konstruktiven Realismus*. Wien: WUV-Universitätsverlag.

Watson, J. D. 1968. *The Double Helix*. New York: Atheneum.

Wegener, A. 1915. *Die Enstehung der Kontinente and Ozeane*. Braunschweig, Germany: F. Vieweg & Sohns (2d ed. 1920, 3d ed. 1922, 4th ed. 1929).

———. 1924. *The Origin of Continents and Oceans*. Tr. J. G. A. Skerl from the 3rd German ed. London: Methuen.

———. 1966. *The Origin of Continents and Oceans*. New York: Dover.

Weinert, F. 1995. *Laws of Nature: Essays on the Philosophical, Scientific, and Historical Dimensions*. New York: Walter de Gruyter.

Westbrook, R. B. 1991. *John Dewey and American Democracy*. Ithaca, NY: Cornell University Press.

Westfall, R. S. 1971. *Force in Newton's Physics*. New York: American Elsevier.

Whitehead, A. N., and B. Russell. 1910–13. *Principia Mathematica*. 3 vols. Cambridge: Cambridge University Press (2d ed. 1925–27).

Whorf, B. L. 1956. *Language, Thought, and Reality: Selected Writings of Benjamin Lee Whorf*. Ed. J. B. Carroll. Cambridge, MA: MIT Press.

Wilson, Daniel J. 1990. *Science, Community, and the Transformation of American Philosophy, 1860–1930*. Chicago: University of Chicago Press.

Wittgenstein, L. 1922. *Tractatus Logico-Philosophicus*. London: Routledge and Kegan Paul.

Woolgar, S. 1981. Discovery: Logic and Sequence in a Scientific Text. In *The Social*

*Process of Scientific Investigation,* Sociology of the Sciences Yearbook, vol. IV, ed. K. D. Knorr, R. Krohn, and R. Whitley, 239–68. Dordrecht: Reidel.

———. 1988a. *Science: The Very Idea.* London: Tavistock.

———, ed. 1988b. *Knowledge and Reflexivity: New Frontiers in the Sociology of Knowledge.* London: Sage.

———. 1991. The Turn to Technology in Social Studies of Science. *Science, Technology, and Human Values* 16:20–50.

Zilsel, E. 1942. The Genesis of the Concept of Physical Law. *The Philosophical Review* 51:245–79.

Ziman, J. 1978. *Reliable Knowledge.* Cambridge: Cambridge University Press.

# Index

acceptance, 185–91
    decision matrix for, 187
    decision rules for, 188–91
    versus belief, 185–86
achromatopsia, 248 n.5
actor networks, 61–63
approximation, 179
Armstrong, D., 86
Ayer, A. J., 220–21

Bacon, F., 123
Bacon, R., 88
Bayesian inference, 156–57
Berlin, B., 100–101
basic color terms, 100–101
    focal colors, 101
Boyle, R., 88, 250 n.8

Campbell, D. T., 46
Carnap, R., 205, 217, 222
    and the *Aufbau,* 219, 259 n.5
    and explication, 156, 223
    and foundationism, 154–56
    and inductive logic, 155–56, 224–26
    and justification, 14
Cartwright, N., 25, 244 n.7, 250 n.16

categories, 100–105
    basic, 103
    in classical mechanics, 106–15
    classical model, 100
    and cognitive models, 105–6
    and exemplars, 104
    graded structure of, 102–6
    natural, 101–5
    and prototypes, 104
    radial structure of, 105–6
    theory-based, 105
causality, 183–85
Chi, M., 112–14
cognitive mechanisms, 48–50
Coffa, A., 90–91
Cohen, I. B., 246 n.13
Cohen, R., 235
Collins, H., 18, 20, 60
concepts. *See* categories
constructive realism, 149–50
    and perspectival realism, 149–50
constructivism, 19–21, 43, 58–60
    epistemological, 19
    ontological, 19
    and realism, 22
correspondence rules, 117, 177

**281**

Cox, A., 136–41, 146
creation science, 245 n.2
crucial decisions, 123–28
  matrix for, 125
  conditions for, 126–25
  decision rule for, 127

Darwin, C., 70–72, 89
Darwinian tale, 196–97
Descartes, R., 87–90
detectability, 181–82
Dewey, J., 12–13, 230–33
  and Trotsky, 233
Doppelt, G., 246 n.7
double helix, 72–75, 191–95, 197–98
discovery vs. Justification, 13–14, 33, 206–7, 227–30

Edinburgh school, 43
Einstein, A., 14, 206–7, 228, 251 n.19
empirical adequacy, 180–81
*Encyclopedia of Unified Science,* 221, 230–31
Enlightenment rationalism, 1, 57–58, 204–5
epistemological norms, 72–75
  as categorical, 72, 75
  as conditional, 72, 74–75
evolutionary models of science, 44–48, 160–63, 171–73
  and norms, 162
experimentation, 52–53
  crucial experiments, 123–24
  *see also* crucial decisions
explanatory coherence, 124
explication, 223–24

Feigl, H., 217, 222, 234, 258 n.2
  and *Readings,* 221
feminist empiricism, 18, 215–16
foundationism, 154–57
Frank, Philipp, 217, 234–35, 244 n.5, 261 n.33

Galileo, 88
gender and science, 17–18
geomagnetic reversals, 136–41
  in deep-sea sediments, 144
  at Jaramillo Creek, 145
  by Juan de Fuca Ridge, 142
  Mammoth event, 140
  Olduvai event, 140
  by Pacific-Antarctic Ridge, 143
  visualizations of, 137, 140, 145
Gilligan, C., 202
Gondwanaland, 129
Gross, P. R., 243 n.1
Grünbaum, A., 234

Hacking, I., 25
  and straight rule, 155
Hanson, N. R., 229
Harding, S., 17, 200
harmonic oscillator, 106–10, 122–23, 175–78
Hempel, C. G., 217, 222–24
  and explication, 223
  and provisos, 90–91
Hess, H., 134–36
  his visual model, 136
Hollinger, D., 244 n.8
Holmes, A., 134
  his visual model, 135
Holton, G., 244 n.5, 261 n.34
Hughes, T., 63–64
Hull, D., 247 n.15

idealized cognitive models, 105–6
incommensurability
  and Kuhn, 38–39
  and Laudan, 42
induction, 196–97, 224–26
  justification of, 196–97
  explanation of, 196–97
inductive logic, 33
interest theory, 43, 61

INDEX | 283

James, W., 71–72
Jeffrey, R., 156, 234
Jewish science, 14, 15, 18, 207, 212, 229
justification
    of scientific knowledge, 31
    of induction, 196–97

Kay, P., 100–101
Keller, E. F., 202–3, 257 n.3
Kepler, J., 88
Köhler, W., 222
Kuhn, T. S., 4, 119–20
    and discovery vs. justification, 151, 208–9
    and incommensurability, 38–39
    and naturalism, 151–52
    reaction to by philosophers, 15–16
    reaction to by sociologists, 43
    on role for history, 151–52, 171–73
    his stage theory, 34–35, 172
    see also *Structure of Scientific Revolutions*
Kyburg, H., 234

Lakatos, I., 16, 152, 209–10
    and metamethodology, 17, 157
    and realism, 78
Lakoff, G., 25, 105–6
Latour, B., 61–63, 254 n.15
Laudan, L., 16, 41–43, 152
    and metamethodology, 157–60
    and naturalism, 17
    and incommensurability, 42
    and realism, 78
laws of nature, 5–6, 23–24
    features of, 86
    history of concept, 87–90
    and necessity, 95–96
    and theology, 249 n.7
    and truth, 90–91
    and principles, 94–95
Levi, I., 234

Logical Empiricism, 32–33, 119, 209
    history of, 13–15
    and pragmatism, 230–33
    as second enlightenment, 57
Longino, H., 202
Longwell, C., 133
Lotze, R. H., 227

MacKenzie, D., 28
man the hunter model, 201–2
maps, 25–26, 81–82, 214–15
McClintock, B., 202–3
McCumber, J., 261 nn. 28, 29
McMullin, E., 16
meaning postulates, 117
mechanics, classical, 106–15, 166–69, 175–77
    basic level models, 110–14
    conservative models, 114–15
    Hamiltonian formulation, 175
    higher-level models, 114–15
    internal structure of models, 115–17
    model map for, 110
Merton, R. K., 19, 33–34, 58
metamethodology, 17, 157–60
*Minnesota Studies*, 221
mobilism, 128
    evidence for, 129–32
modality, 182–85
model map, 110
models, 72–73, 91–94, 175–77
    basic level of, 110–14
    in classical mechanics, 106–15
    complexity of, 109
    and concepts, 99–100
    and generalizations, 94
    and hypotheses, 177–82
    and language, 123
    and prototypes, 122–23
    radial structure of, 105–6
    and theories, 122–23, 165–69
    and the world, 123
    visual, 110, 131–32

Morris, C. W., 230–31
Moore, G. E., 223

Nagel, E., 226, 230, 234, 261n.32
National Socialism, 13, 15, 229
naturalism, 5, 53–54
   arguments against, 153–54
   and explanation of norms, 72–75
   metaphysical, 70
   methodological, 69–70
   priority, 70–72
   and realism, 60–61, 77–79
   theoretical, 77
naturalization, 149
Newton, I., 87–90
   use of diagrams, 118
   conception of God, 249n.6
Nietzsche, F., 4
novice-expert shift, 107, 113–14, 252n.13

observation, 79–81
one-world rule, 82–83

paradoxes of confirmation, 224
paradigm, 36–37
pendulum, 106–10, 122–23
perspectival realism, 26, 79
perspectivism
   and observation, 79–81
   and theories, 81–82
philosophy of science
   autonomy of, 28
   historical school, 16–17
   history of, 11–18
   and science, 30–32
   and science studies, 29
Pickering, A., 2, 21, 60, 248n.13
Peirce, C. S., 75, 232
Pinch, T., 18, 60
Plantinga, A., 248n.2
Popper, K. R., 211
   and foundationism, 155
   and justification, 14
   and world three, 46

postmodernism, 58–60
pragmatism
   and naturalism, 75–77
   and Logical Empiricism, 230–33
   and McCarthyism, 233
probability
   and Carnap, 225
   and induction, 224–26
   model-based, 127
   as propensity, 127
   and Reichenbach, 226
psychologism, 33
Putnam, H., 233
   argument against naturalism, 163–65
   refutation of realism, 24

Quine, W. V. O., 152, 154, 183, 230, 234

rationality, 26–28
realism, 21
   constructive, 149–50, 168, 180–82
   and constructivism, 22
   feminist, 215–16
   metaphysical, 163
   and modality, 95–96, 183–85
   and naturalism, 60–61
   perspectival, 26, 150, 212–15, 240–41
   and reflexivity, 21
   and truth, 6
   as vision of science, 22–26
reflexivity, 20–21
   and realism, 21
Reichenbach, H., 13–14, 205–8, 219–20, 222, 259n.17, 261n.30
   appointment in Berlin, 13, 258n.3
   and Dewey, 260n.24
   and discovery vs. justification, 13–14, 206–7, 227–30
   and foundationism, 155
   and induction, 14, 155, 226, 260n.23

and *Experience and Prediction,* 13, 206–7, 220, 227–28
representation, 50–51, 213–15
Rescher, N., 234
Rosch, E., 101–5, 111–12, 252 n.13
Ross, A., 243 n.1

Salmon, W., 233
satisficing, 190–91
Schlick, M., 217, 220
science
   and philosophy of science, 30–32
   success of, 31
science studies
   autonomy of, 29
   goals of, 22–23
   as multidisciplinary, 3, 63
   and philosophy of science, 29
   vision of, 28–29
science wars, 1–8
scientific knowledge, justifiability of, 31
scientific philosophy, 12–13
sea-floor spreading, 134–46
   Holmes' model, 134–35
   Hess' model, 135–36
   Vine-Matthews hypothesis, 137–39
semantic view of theories. *See* theories, model-based view of
Schlick, M., 205, 220
Shapere, D., 16, 152, 247 n.2
Simon, H., 254 n.17
sociology of science
   constructivism, 18–21, 43, 58–60, 120–21
   functionalism, 18–19, 33–34, 58
Sokal, A., 2
stabilism, 128
*Structure of Scientific Revolutions* (Kuhn), 34–40, 235, 245 n.5
   cognitive aspects of, 39–40
   evolutionary aspects of, 40
   pre-paradigm science, 35

normal science, 36–37
crisis, 37
revolution, 37–38
*see also* Kuhn, T. S.
Suppes, P., 176, 234

technology
   policy, 64–65
   social shaping of, 64
theories
   as maps, 25–26, 81–82
   model-based view of, 72–73, 98–99, 122–23, 165–69, 175–82, 251 n.1
   as perspectival, 81–82
   as propositional, 124
   received view of, 97–98
   *see also* models
theory choice, 51–52, 124–25, 169–71
   model-based, 127
theory of science, 197–99, 257 n.20
Toulmin, S., 16, 46, 152

underdetermination, 237–41
   deductive, 238–39
   methodological, 240
   and social constructivism, 237–38

van Fraassen, B. C., 150, 174–99, 209, 256 n.1, 262 n.35
Vienna Circle manifesto, 218
Vine, F., 135–141
Vine-Matthews hypothesis, 137–39

Wartofsky, M., 235
Watson, J. D., 73–75, 191–95
Wegener, A., 128–34
Wilson, J. T., 141
Wittgenstein, L., 220, 223
woman the gatherer model, 201–2
Woolgar, S., 20, 59–60